ISSUES IN TEACHING MATHEMATICS

Other titles in the Cassell Education series:

P. Ainley: *Young People Leaving Home*

P. Ainley and M. Corney: *Training for the Future: The Rise and Fall of the Manpower Services Commission*

G. Allen and I. Martin (eds): *Education and Community: The Politics of Practice*

G. Antonouris and J. Wilson: *Equal Opportunities in Schools: New Dimensions in Topic Work*

M. Barber: *Education in the Capital*

L. Bash and D. Coulby: *The Education Reform Act: Competition and Control*

D. E. Bland: *Managing Higher Education*

M. Booth, J. Furlong and M. Wilkin: *Partnership in Initial Teacher Training*

M. Bottery: *The Morality of the School*

G. Claxton: *Being a Teacher: A Positive Approach to Change and Stress*

G. Claxton: *Teaching to Learn: A Direction for Education*

D. Coffey: *Schools and Work: Developments in Vocational Education*

D. Coulby and L. Bash: *Contradiction and Conflict: The 1988 Education Act in Action*

D. Coulby and S. Ward (eds): *The Primary Core National Curriculum*

L. B. Curzon: *Teaching in Further Education* (4th edition)

P. Daunt: *Meeting Disability: A European Response*

J. Freeman: *Gifted Children Growing Up*

J. Lynch: *Education for Citizenship in a Multicultural Society*

J. Nias, G. Southworth and R. Yeomans: *Staff Relationships in the Primary School*

A. Pollard and S. Tann: *Reflective Teaching in the Primary Schools* (2nd edition)

R. Ritchie (ed.): *Profiling in Primary Schools: A Handbook for Teachers*

A. Rogers: *Adults Learning for Development*

B. Spiecker and R. Straughan (eds): *Freedom and Indoctrination in Education: International Perspectives*

A. Stables: *An Approach to English*

R. Straughan: *Beliefs, Behaviour and Education*

M. Styles, E. Bearne and V. Watson (eds): *After Alice: Exploring Children's Literature*

S. Tann: *Developing Language in the Primary Classroom*

H. Thomas: *Education Costs and Performance*

H. Thomas with G. Kirkpatrick and E. Nicholson: *Financial Delegation and the Local Management of Schools*

D. Thyer: *Mathematical Enrichment Exercises: A Teacher's Guide*

D. Thyer and J. Maggs: *Teaching Mathematics to Young Children* (3rd edition)

P. Walsh: *Education and Meaning: Philosophy in Practice*

M. Watts: *The Science of Problem-Solving*

M. Watts (ed.): *Science in the National Curriculum*

J. Wilson: *A New Introduction to Moral Education*

S. Wolfendale *et al.* (eds): *The Profession and Practice of Educational Psychology: Future Directions*

Issues in Teaching Mathematics

Edited by

Anthony Orton and Geoffrey Wain

CASSELL

Cassell
Wellington House
125 Strand
London WC2R OBB

PO Box 605
Herndon
VA 20172

First published 1994
Reprinted 1996 .

British Library Cataloguing-in-Publication Data
A catalogue record for this book is available from the British Library.

ISBN: 0-304-32678-X (hardback)
      0-304-32680-1 (paperback)

Typeset by Colset Private Limited, Singapore
Printed and bound in Great Britain by Redwood Books, Trowbridge, Wiltshire

Contents

About the Contributors

David Carter is a Lecturer in Mathematical Education at the University of Leeds, having previously held posts as a secondary school mathematics teacher and head of department for twelve years. He has served as Chair of the Teaching Committee of the Mathematical Association. His special interests include the assessment of mathematical achievement and attainment and students' understanding of motion in a circle. He has also been a mathematics teacher training consultant in Kenya.

Leonard Frobisher is the Mathematics Coordinator with the Mathematics at Key Stage 2 Assessment Project. He has previously worked both on the evaluation of the National Curriculum Project and as a Lecturer in Mathematical Education in the School of Education at the University of Leeds and at Leeds Polytechnic. His interests are in investigative approaches to learning mathematics and innovative ways of assessing achievement and attainment.

William Gibbs has worked as a mathematics educator and consultant in many parts of the world, including Bhutan, India, Sierra Leone, Solomon Islands and Zambia. His research interests centre on the relationship of mathematics to the community it serves and the use of simple resources in teaching mathematics. Recently he has been a Lecturer in Mathematical Education in the School of Education at the University of Leeds. He is co-author of *Health into Mathematics*.

John Monaghan is currently a Lecturer in Mathematical Education at the University of Nottingham. He previously held a similar post at the University of Leeds, after being a school teacher of mathematics and department head for many years. He has research interests in pupils' learning, new technologies in education and teacher education. He is co-editor of *Computer Algebra Systems in the Classroom*.

Anthony Orton is a Senior Lecturer in Mathematical Education in the School of Education at the University of Leeds. He was a secondary school teacher of mathematics and department head for ten years before moving to teacher education at a College of Higher

Education. He has been actively involved in research in learning mathematics for over twenty years and has been involved in developmental work overseas for ten years. He is the author of *Learning Mathematics; Issues, Theory and Classroom Practice.*

Jean Orton works as a part-time Lecturer, mainly with overseas students, and as a part-time research assistant in Mathematical Education in the School of Education at the University of Leeds. She taught mathematics for a number of years in secondary schools and has worked in both primary and pre-school education. She has acted as a mathematics consultant overseas, most recently in Sierra Leone. Her current research interests are in children's perception and use of pattern.

Tom Roper is a Lecturer in Mathematical Education in the School of Education at the University of Leeds, having previously worked as a secondary school teacher and department head for 16 years. He has contributed to the Mechanics in Action Project since its inception and has been co-author of a number of texts on mechanics at school and undergraduate levels. His current research interests are in the assessment of Mal and the use of computer algebra systems.

Geoffrey Wain is a Senior Lecturer in Mathematical Education in the School of Education at the University of Leeds. He was co-director of the Mathematics Teacher Education Project and has been Chairman of the Association of Mathematics Education Tutors and Secretary of the Joint Mathematical Council of the United Kingdom. He has been extensively involved in developmental work overseas and has directed a project on using home-owned microcomputers in education. His current research interests include computer algebra systems.

Preface

This book aims to provide the reader with background knowledge and understanding of some major contemporary issues facing mathematics teachers. In the different regions of Britain we now have, for the first time, a centrally prescribed curriculum, though many other countries have been in this position for years. Within the inevitable constraints imposed by such a curriculum, it ought to be possible to take account of new branches of mathematics, new approaches to teaching, new means of assessment and evaluation, and ever more sophisticated technology which cannot be ignored. Much of the book is devoted to discussion of such issues in relation to curriculum construction and change. The social and psychological contexts of mathematics teaching always command a place, for in both of these areas research is continually providing new knowledge and ideas which teachers need to evaluate for themselves. Issues concerned with the relationship between language and mathematics are also continually being enlightened by research, and this topic is dealt with in a separate chapter, as is the important issue of mathematics across the curriculum. In short, our aim has been to enable mathematics teachers to become better informed across a range of issues which have implications for now and for the immediate future, and thus better able to reflect on the wider concerns which underlie and inform the day-to-day activities of the mathematics classroom. We hope we have achieved our aim of producing a relatively short book which nevertheless considers the major issues of today in sufficient depth to be coherent.

Tony Orton
Geoff Wain
Leeds, 1993

Abbreviations

ACACE	Advisory Council for Adult and Continuing Education
A-level	Advanced level GCE
APU	Assessment of Performance Unit
AT	Attainment Target (of the National Curriculum)
ATCDE	Association of Teachers in Colleges and Departments of Education
ATM	Association of Teachers of Mathematics
CAS	Computer algebra system
CGS	Computer graphics system
CNAA	Council for National Academic Awards
CSE	Certificate of Secondary Education
CSMS	Concepts in Secondary Mathematics and Science
DES	Department of Education and Science, now the DFE
DFE	Department for Education, formerly the DES
ENCA1	Evaluation of National Curriculum Assessment at KS1
ERA	Education Reform Act, 1988
GAIM	Graded Assessment in Mathematics
GCE	General Certificate of Education
GCSE	General Certificate of Secondary Education
HE	Higher Education
HMI	Her Majesty's Inspector(s)
ICMI	International Commission on Mathematical Instruction
ILEA	Inner London Education Authority
INSET	In-service Education and Training
IT	Information technology
JMB	Joint Matriculation Board (see also NEAB)
KMP	Kent Mathematics Project
KS1, 2, 3, 4	Key Stages of the National Curriculum
LEA	Local education authority
LMS	Local management of schools

Ma1, 2, 3, 4, 5	The five Attainment Targets of the National Curriculum (Mathematics)
MA	Mathematical Association
MAG	Mathematics Applicable Group
MAP	Mechanics in Action Project
MEI	Mathematics in Education and Industry
MME	Midlands Mathematical Experiment
NEAB	Northern Examinations and Assessment Board
NC	National Curriculum of England and Wales
NCC	National Curriculum Council
NCTM	National Council of Teachers of Mathematics
PrIME	Primary Initiatives in Mathematics Education
PoS	Programme of Study
O-level	Ordinary level GCE
SAT	Standard Attainment Test
SCUE	Standing Conference on University Entrance
SEAC	School Examinations and Assessment Council
SEC	Secondary Examinations Council
SM	Symbol manipulator
SMILE	Secondary Mathematics Individualized Learning Environment
SMP	School Mathematics Project
SoA	Statement of Attainment
TGAT	Task Group on Assessment and Testing
TVEI	Technical and Vocational Education Initiative
UODLE	University of Oxford Delegacy of Local Examinations

Chapter 1

The Aims of Teaching Mathematics

Anthony Orton

THE IMPORTANCE OF AIMS

> There are important aims which should be an essential part of any general statement of intent for the teaching of mathematics.
> (HMI, 1985a)

In the above statement from Her Majesty's Inspectors we are reminded of the long-standing belief that aims are important in education, that they not only form a part but perhaps come first in the definition of a curriculum. Ever since the thoughts of educators first began to be remembered and retained for posterity, the records of their statements have been liberally sprinkled with discussion of the purposes of education and their relationship with curriculum construction. Thus the curricula of many countries around the world today are prefaced by statements of aims, or goals, or objectives, or some combination. Yet the most recent National Curriculum documents for mathematics teachers in England (see Chapter 4) contain little or nothing in the way of explicit statements of aims. Does this indicate that we now have an aimless mathematics curriculum? Is it that so much has been written in the past about the aims of teaching mathematics that all is now agreed and so it is no longer necessary to consider the issue in the preparation of a curriculum? If we believe that aims are now firmly established in the minds of all educators and learners alike, and if we believe that aims remain constant over time, then the answer to the question might well be in the affirmative. If we believe that teachers and learners do not need to think about aims, they simply need to be told what to do, then we might likewise consider that the discussion of aims was unnecessary. If, however, we believe that aims can change with the passage of time, or if we believe that teachers do need to reflect on the purposes of the undertaking in which they are fundamental contributors, then a consideration of aims must remain an ongoing theme within the thinking and planning of all teachers. Bruner (1966) said, 'I shall take it as self-evident that each generation must define afresh the nature, direction, and aims of education.' The National Council of Teachers of Mathematics (NCTM, 1989) declared that 'Calls for reform in school mathematics suggest that new goals are needed.' Alexander, Rose and Woodhead (1992) indicated that 'The teacher must be clear about the goals of learning before deciding on methods of organisation.' Sosniak, Ethington and Varelas (1991) have produced evidence that 'eighth grade mathematics teachers . . . apparently teach their subject *without* a theoretically coherent point of view', with the implication that this state of affairs is undesirable. That aims need to be kept under constant review is clearly essential.

There are many reasons why we need to keep our aims under constant review. One reason is that what we and others see taking place in our classrooms does not always, on careful reflection, seem justifiable. This in turn might make us wonder whether we ourselves ever really grasped what our purposes should be. Another is that pupils themselves do, from time to time, express concern about the end-points of their current studies, though it must be admitted that these concerns are sometimes the outcome of short-term frustrations with their ongoing work. A third reason is that situations and circumstances do change, new content and new teaching methods are proposed, and any proposal for change should automatically force us to rethink. Our aims underlie all such questions and concerns. Aims are fundamental because without them we may not know where we are going or why, we may drift and wander through the immense field of mathematics without clear ends in view. Even when we deliberately allow lessons to wander, in the sense of developing according to whatever ideas are put forward by the learners, there should still be purpose behind what we do. The educational process should be constantly guided by aims. When we consider what is actually being taught in classrooms today we are likely to find many items which urgently demand rejustification. Any justification is bound to reflect our aims, and indeed perhaps reflect on whether we have thought about our aims.

One example of an item within mathematics which requires rejustification is long division. Given that long division using paper and pencil is a difficult process which many children cannot complete without error and which even fewer understand, and given that modern technology has provided us with alternative ways of carrying out long divisions, it would seem appropriate to go back and reconsider our aims in respect of this algorithm. What we mean by 'understand' will be considered more fully in Chapter 3. Suffice it to say at this point that we certainly mean more than 'know the procedure', we also include comprehension of why the procedure works and when it is appropriate to use it. Watson (1992) has compared the current situation with regard to the long-division algorithm with other algorithms which have been removed from the curriculum, in particular the square-root algorithm. Some readers will remember that the square-root algorithm was once taught alongside the use of tables, but more recently the algorithm has been omitted from the curriculum and, for a time, tables alone were used. Nowadays we seem to have accepted the calculator as the obvious means by which to determine square roots. What is more, the calculator is so versatile that roots of all kinds can now be obtained very easily. Supporters of the use of the calculator might well argue that, in a society which uses automatic washing machines and spin dryers, it makes no sense to insist on all children learning to use a mangle, the old hand-operated device for squeezing excess water out of clothing and linen. It does make sense, however, for society to store away somewhere the knowledge of what a mangle can do and how to make one. The NCTM (1989) wrote, 'Some calculations, if not too complex, should be solved by following standard paper-and-pencil algorithms . . . For more complex calculations, the calculator should be used (column addition, long division).' The calculator is now ubiquitous in many societies, and there is no evidence that children make more errors when using a calculator than they ever did with pencil and paper. The algorithm can be best understood through simple short divisions; the algorithm can best be carried out by means of the calculator. Furthermore, the pencil-and-paper algorithm is very much a product of Western mathematics. The abacus has long provided the means of carrying out arithmetical processes in the Eastern world. As long ago as 1938 Spens wrote, 'Tools become obsolete and better ones take their place', and 'There is little profit in spending much time in perfecting the command of a tool which will be rarely used in later years, and it is a mistake to delay the introduction of the newer and better tool which has replaced it.' So what sort of aims might lead to an insistence on teaching long division using pencil and paper nowadays? Would a review and consequent restatement of aims assist society to decide on what is the proper place for calculators in school mathematics?

Similar arguments might apply to the topic of fractions, which runs through the whole of school mathematics, starting with basic ideas of fractions, working through comparison of size, addition, subtraction and multiplication, and culminating in division. For some pupils, rational numbers form a very important part of the study of arithmetic and algebra, and for many pupils a satisfactory level of proficiency in handling fractions might still be important. But many other children study the same fraction algorithms year after year without appreciably improving their performance and without retaining procedures from one year to the next. All the evidence which exists suggests that ratio and proportion are very difficult concept areas for children, and the fraction algorithms to some extent depend on understanding ratio and proportion. There is evidence that most adults, even those educated through older, more 'traditional' curricula (Sewell, 1981), are not able to handle simple algorithms such as those concerning fractions, so this deficiency cannot be simply attributed to modern teaching. Once again, there are calculating devices which will handle the processes. There are relatively cheap scientific calculators through which arithmetic fractions may be manipulated, and there are now more sophisticated calculators and hand-held computers which allow similar operations on algebraic fractions (see Chapter 12). At the present moment, the world of mathematics teaching is having to begin to come to terms with computer algebra systems and graphic calculators in the same way that it had to begin to accommodate pocket calculators some time ago. This process of accommodation is difficult unless there is some measure of agreement about aims.

There is no intention, here, of simply arguing the merits and demerits of particular chunks of curriculum content in any detail. Examples have been given which might lead to debate in which particular views ought to be based on what we are setting out to achieve, on what end-points we have in mind for our children. If we clarify and agree on these issues, curriculum detail should then follow, though there is still likely to be some choice. Contemporary curricula around the world are sometimes said to be based on a blend of traditional and modern, these terms being perhaps simplistically described as pre-1960s and immediate post-1960s. Were the particular topics, either traditional or modern, chosen to satisfy particular predefined aims, or were they chosen merely to achieve an acceptable blend? Of course, a curriculum does not only consist of content. It might be that particular methods of teaching might achieve particular aims rather than content. But, for example, are we clear what we are setting out to achieve when we insist that children should carry out open-ended investigations which do not produce significant content or factual knowledge, but which allow children to be assessed on process skills which might begin with describing what they did, what they discovered and how far they got?

One worthy aim for teachers of mathematics is to attempt to keep learners constantly attentive and motivated, to enable pupils to come to believe that the subject is relevant and worth studying. Learners of all ages can quickly become demotivated if they perceive the curriculum to be dull and boring, if their interpretation of what is expected of them is that it is without real purpose, has no relevance to reality as they see it, or is too hard and unintelligible. In other words, children will deduce their own views of the purposes of learning mathematics and of the aims which underlie particular aspects of the curriculum, and we might be horrified by their deductions. The 'hidden' curriculum is a phenomenon which has been much discussed (see for example Layton, 1978), and part of that hidden curriculum is likely to be a perception of our aims. What do children think are the aims of learning mathematics? The worry is that they might deduce some, or all, of the following:

- to ensure that learners find mathematics is irrelevant to their world, the only real world they know;
- to contribute to the view that school learning is dull and dreary;
- to maximize the number of people who find mathematics unintelligible;
- to minimize the number of people who learn to love and enjoy mathematics;
- to minimize the number of qualified mathematicians in society.

Sometimes the opinion is expressed that, since life is not a bed of roses, it is therefore a good idea for children to come to terms with its reality while at school. Such an attitude is presumably itself an expression of an educational aim, namely that children should be prepared for the drudgery of life. Many others would reject what they would regard as a negative attitude and in doing so would wish to do all they could to prevent such a hidden set of aims from becoming formed. In order to try to prevent such aims being formed by our pupils, we need to be sure of what our aims really are. But would we all come to the same conclusions?

AIMS, GOALS AND OBJECTIVES

What are aims, what are goals, and what are objectives? Does differentiation of meaning really matter? The literature on education seems to be liberally sprinkled with all three words, while other words like 'intentions' and 'purposes' do not appear to have been accorded the status of technical terms in quite the same way. Is there any difference of meaning between any of these words? Dictionaries, which only state common usage, certainly do not appear to differentiate, and it would perhaps be unreasonable to expect them to contain all the specialist uses of words, including those specific to education. In fact, dictionaries are likely to explain one of the words by referring to some of the others, so if there are distinctions of meaning within education it would seem that these distinctions have been created by education.

It is convenient to attempt a differentiation of meaning as used within education by starting with the word 'objective'. Some years ago, a great deal of discussion took place within education on the concept of 'behavioural objectives'. A behavioural object-ive was considered to be a statement of intent which described the expected change in the behaviour of the learner as a result of being exposed to the prescribed learning experiences. Thus, 'the pupils will be able to add together two fractions' seems a clear objective which specifies a capability that learners are expected to acquire by the end of the period of teaching, and that they presumably could not demonstrate at the start. In this way, the potential behaviour of the individual was to be changed. Such instruc-tional objectives have been the subject of detailed analysis by, for example, Mager (1975). He was at pains to separate the concept of an instructional objective from behaviourism, which is basically a theory of how learning takes place, while other writers, for example Gagné (1970), have implied that objectives were very much a part of behaviourist views on learning. Such debate was a particular feature of the 1970s. Nowadays we are likely to regard the idea of objectives as too valuable a one to become lost to good educational practice through being closely associated with a learning theory to which few educationists now wholeheartedly subscribe. It is considered by many that precisely defined objectives can serve a number of purposes. They can, for example, provide teachers with guidelines for the planning of teaching sequences and methods; thus teacher training typically encourages trainee teachers to begin the pro-cess of planning a lesson by defining clear objectives. In the assessment of achieve-ment, it is also very helpful if clearly defined objectives can be used as a benchmark against which performance can be measured, and so examination syllabuses may well be prefaced by statements of objectives. Furthermore, learners themselves deserve to know the objectives of a unit of study in order the better to motivate themselves, to direct their efforts and to evaluate their own progress continually. Clearly defined objectives still seem to play an important part in systematic education, and this is revealed in HMI (1979), where the relating of objectives with 'milestones along the path of progression and development' appears to anticipate the Attainment Targets of the National Curriculum of England and Wales. Some teachers might now say the milestones have unfortunately become millstones round their necks!

The difficulty with objectives arises from their very nature and value. If they are to serve a useful purpose they need to be totally specific and beyond variation of inter-pretation. For much of the time this is well nigh impossible. Earlier, an example of objectives was provided, concerning the addition of two fractions. But what do we mean by fractions? Do we mean only proper fractions or are improper fractions included as well? If so, are they included in their mixed number form? And how com-plex should the fractions be? Are we interested only in single-digit numerators and/or denominators? Do we expect answers to be reduced to their simplest form when appropriate? Thus 'the derivation of complete, detailed, unambiguous and absolutely specific objectives is itself an elusive objective to have' (Orton, 1992).

Thus far, very little has been said about aims, the focus of this chapter. An element-ary definition of aims is to say that they describe our purposes, intentions or end-points without attempting to be specific in the way that objectives set out to be. Thus, having considered objectives makes it possible to describe the idea of aims through com-parison. Aims are less specific and more general. Examples of aims might include 'to develop proficiency in combining fractions'. An even more general aim is 'to develop skill in number work'. An aim such as 'to improve mathematical performance' would probably be considered by most educators to be too vague and general to be helpful, and yet 'to improve attitude to mathematics' probably would not! Thus our definition of aims has to be seen in contrast to the greater precision expected from objectives. In essence, and to some people, there may be little difference, in that both express intentions, purposes and end-points, but the difference may lie in where they are situated on a spectrum from generality to specificity. The working definition assumed here is that aims are broad statements of the desired outcomes of the education process while objectives are more specific and detailed and may relate to an extremely small and limited part of a curriculum. So what, then, are goals? The NCTM, for example NCTM (1989), often appears to use the word 'goals' as synonymous with aims. Scopes (1973) used all three words, 'aims', 'goals' and 'objectives', without attempting to suggest that common usage in education might wish to distinguish between any of them. Perhaps, to some people, goals relate to the aims of education while objectives refer to mathematics education, but otherwise it is difficult to see any other differentiation of meaning. What it amounts to is that different writers always have used and always will use the same words in different ways. For the purposes of reading this book with understanding, an attempt has been made to distinguish between aims and objectives, but no attempt will be made to give a specific and different meaning to 'goals'.

PRIMARY AND SECONDARY AIMS

Aims can be viewed as being of a variety of kinds and perhaps even at a variety of levels. One sort of level is associated with the setting of aims for mathematics teaching within the context of overall educational aims. General educational aims do, presumably, come first, and our aims for teaching mathematics should then follow as a logical consequence. The aims for teaching mathematics should not be in conflict with our general educational aims and should, indeed, ideally contribute to the attainment of one or more of the general aims. Furthermore, our aims must be such that mathematics is accepted as an essential curriculum component, as offering to the overall curriculum what cannot be obtained from any other content area which is perhaps clamouring for a place in the overall curriculum. The relationship between general educational aims and aims of mathematics teaching is, perhaps, one indication of the fact that there might be various levels of aims. But other concepts of primacy or levels of aims exist. Spens (1938), for example, wrote, 'No school subject, except perhaps classics, has suffered more than mathematics from the tendency to stress secondary rather than primary aims, and to emphasize extraneous rather than intrinsic values', indicating that debate on primacy has been engaged before.

An obvious concept of levels of aims is associated with the fact that, first and foremost, we have to have reasons for including mathematics in the school curriculum. Such aims have the right to be referred to as primary aims. Why is the particular domain of the totality of knowledge labelled 'mathematics' worth a place when many others are not, for example law, politics and psychology? Secondly, once it has been decided that mathematics must be included within the school curriculum, the detail of the content and methods might be based on secondary aims such as that pupils should appreciate that there are different kinds of geometries, or that pupils should understand how society in the past has had to perform calculations without the benefit of modern technology. Thirdly, there might be important general aims of education which can be pursued within mathematics once it has been decided to teach mathematics but which could equally well be learned through other subject matter and content. Children nowadays frequently work together in small groups, and this might be for a variety of reasons, one of which might be to encourage cooperative working. Such cooperative working can indeed be pursued within mathematics, but is it really one of the primary aims of teaching mathematics? If one of our aims of education is to develop cooperative learning abilities in pupils, must it be a part of all subjects on the curriculum? Such an aim could hardly be considered a primary aim of mathematical education, but might be a primary aim of education.

An interesting mixed bag of aims is to be found in HMI (1985a). The headings for the ten aims, which are subsequently discussed in some detail, are:

- mathematics as an essential element of communication;
- mathematics as a powerful tool;
- appreciation of relationships within mathematics;
- awareness of the fascination of mathematics;
- imagination, initiative and flexibility of mind in mathematics;
- working in a systematic way;
- working independently;
- working cooperatively;
- in-depth study in mathematics;
- pupils' confidence in their mathematical abilities.

Here, there appear to be two aims which are reasons for teaching mathematics but which few would consider to comprise a comprehensive statement of primary aims, namely mathematics as communication and as a powerful tool. Also, there are three aims which seem to be general aims of education which the writers feel should be manifested in mathematics lessons, namely working systematically, independently and cooperatively. The remaining ones are a selection of mathematical aims which could be incorporated once it has been decided to teach mathematics. While such a list of aims, and particularly the accompanying discussion, should be of great value to many teachers of mathematics, it is a pity that a more structured approach was not adopted. The discussion in HMI (1979) conveys a clearer message about levels of aims than their more recent document and, in fact, describes the reasons for teaching mathematics not as aims but as purposes.

A further hierarchy of aims could be based on who devises them. Professional educators would certainly expect to be held responsible for the provision of a statement of aims, but one point of difficulty which then immediately arises is that aims devised by educators might not be deemed wholly acceptable to parents, to prospective employers, or to other groups in society who might feel that their views are important and have not been sufficiently taken into account. Thus it is that government might feel obliged to play a part in the definition of the aims of teaching mathematics, claiming to act on behalf of parents or society as a whole, or even the perceived needs of the state. Pupils, too, will have their own views on the aims which underpin their learning of mathematics, in terms of what they wish it to do for them. It would not be unusual for the other groups to dismiss aims defined by pupils as being naively constructed and not benefiting from the wider perspectives on life which only increased maturity can bring. Thus there could be another hierarchy of aims, but which group should we then regard as providing the primary aims? And whose aims take precedence when there is conflict? In view of the multiplicity of kinds and levels of aims, the remainder of this chapter is devoted to a consideration of the primary reasons for the place which mathematics holds in school curricula around the world, and is set against the background consideration of the aims of education.

THE AIMS OF EDUCATION

It would perhaps be naive to suggest that the aims of mathematical education should always be directly and solely developed from the aims of education. Aims evolve and change in response to trends in society, and these trends, and perhaps even pressures, might be related only to mathematics. However, the aims of mathematical education should ideally evolve with considered acknowledgement to the prevailing aims of education, and cannot sensibly develop in opposition to these more general aims. Educational aims are, themselves, likely to be derived in accordance with particular beliefs, of course, and values and philosophical views held by individuals are rarely agreed by all. Spens (1938) wrote, 'any educational aims which are concrete enough to give definite guidance are correlative to ideals of life – and as ideals of life are eternally at variance, their conflict will be reflected in educational theories.' The NCTM (1989) wrote, 'the goals all schools try to achieve are both a reflection of the needs of society and the needs of students.' This 'official' view seems hardly to have changed when one considers the following statement from Spens (1938):

> school . . . is to be regarded as not merely a 'place of learning' but as a social unit or society . . . it is deliberately created and maintained as a means of bringing to bear upon the young formative influences deemed to be of high importance either for their own development or for the continued well-being of the community.

The view that both the individual and society (or the state) are joint beneficiaries of education seems to be commonly held. For example, in curriculum documents issued to Zambian teachers we have: 'To equip the child to live effectively in this modern age of science and technology and enable him/her to contribute to the social and economic development of Zambia', and in similar documents from Cameroon: 'The structure . . . is based on important national needs and the needs for the general culture of modern man, no matter what his profession or specialism will be.'

It should be clear that particular beliefs might therefore be influenced by political opinion or ideology. One possible particular point of disagreement is to what extent education is for the benefit of the individual alone and to what extent it is for what the individual can do for society or for the state. The two extreme views are that, on the one hand, education is solely for the benefit of the individual who has no subsequent obligations to society in a wider sense, and, on the other hand, since it is the state which finances the education of the individual, then the education received should be geared so that the state is the first beneficiary. School education is never financed by the child but, looking across the whole world, is sometimes at least partially financed by the parents. And in any case, it is income from taxation which finances the state. Under such circumstances it would seem legitimate for the benefits to the child to be considered first. Should education even be for how the individual might be able to change society for the better? Here we are certainly on political ground. Change which is considered by some to be for the better might be considered undesirable by others. It seems unlikely, therefore, that we can be categorical about the aims of education, which is not the most helpful of starting points for a consideration of aims in the context of mathematics teaching. We have to accept that there might be conflicting opinions, but at least we are at liberty to come to our own views on the matter. In doing so, however, it would seem reasonable to look at the ideas of others who by their status in the world of education demand attention.

Even then, it is hard to know what to select from the many. Whitehead (1932) is much quoted and, in declaring that 'Education is the acquisition of the art of the utilisation of knowledge', shows he was clearly concerned to take the emphasis away from the acquisition of knowledge alone. Some knowledge is clearly important, but perhaps it is much more important to know not only how one might use that knowledge, in an instrumental sense, but also to have an awareness of the possibilities and implications of using knowledge in particular ways. Whitehead was a mathematician in the first instance, and if we are to base our aims of mathematical education only on aims of education as expressed by a mathematician we may lay ourselves open to criticism. Warnock (1978) gave a much more extensive definition in saying:

> The goals [of education] . . . are, first, to enlarge a child's knowledge, experience and imaginative understanding, and thus his awareness of moral values and capacity for enjoyment; and secondly, to enable him to enter the world after formal education is over as an active participant in society and a responsible contributor to it, capable of achieving as much independence as possible.

An even more detailed statement comes from HMI (1985b):

- to help pupils to develop lively, enquiring minds, the ability to question and argue rationally and to apply themselves to tasks, and physical skills;
- to help pupils to acquire knowledge and skills relevant to adult life and employment in a fast changing world;
- to help pupils to use language and number effectively;
- to instil respect for religious and moral values, and tolerance of other races, religions and ways of life;
- to help pupils to understand the world in which they live, and the interdependence of individuals, groups and nations;
- to help pupils to appreciate human achievements and aspirations.

Most countries have aims of education. Here, for comparison, are the aims (called goals) drawn up for Malaysia (Curriculum Development Centre, Malaysia, 1975):

- to nurture a balanced development in each individual by providing for the growth of physical, intellectual, emotional, moral and aesthetic potentials as a Malaysian upholding the tenets of Rukunegara ['National Principles'];
- to assist the individual to obtain greater insights and understanding into our ecological and cultural heritage, social institutions, values and practices, societal pressures and challenges, to enable the individual to function and fulfil his commitments and responsibilities as a citizen;

- to develop the human resources of the nation by assisting the individual to be a skilled, competent, rational and responsible planner, producer and consumer, to enable him to improve his personal well-being and contribute to the progress and development of the nation;
- to develop in the individual understanding and acceptance of the democratic ideas and ideals under the Constitution, loyalty to the King, patriotism to the nation, awareness of the rights and responsibilities as a citizen in a democracy and commitment to exercise these rights and responsibilities;
- to develop in the individual positive attitude towards scientific enquiry and technical processes and progress, self-reliance, desire and capability for life-long education to enable him to initiate and adapt to changes compatible with the cultural and ethnical values and aspirations of the nation.

From guidelines issued to teachers in Swaziland, however, we have, 'The goal of basic education is to ensure, as far as possible, that pupils leaving school . . . mature into responsible, self-reliant, productive and happy citizens.' The reference to happiness is a welcome but unfortunately rare inclusion in statements of educational aims.

Many more examples could be given, but there is a sufficient number and variety to enable the reader to reflect on how these various statements of aims might lead directly or indirectly to aims for teaching mathematics. A more difficult decision is whether we are entitled to define aims for mathematics learning which do not relate to any expressions of general aims for education.

THE NATURE OF MATHEMATICS

It has already been claimed that aims should be a precursor to the detailed specification of a curriculum, but that the aims of education are inevitably themselves based on beliefs and attitudes. As regards mathematical education, one important underlying consideration is what one believes mathematics to be. Unfortunately, it seems as if there are almost as many views of what mathematics is as there have been mathematicians. Dictionary definitions include 'the study of number, form, arrangement and associated relationships using clearly defined literal, numerical and operational symbols', and 'science of space and number'. 'Mathematics is what mathematicians do' is much quoted as a definition (see for example Bell, 1952), though its origins are obscure. Russell (1921) stated that 'pure mathematics may be defined as the subject in which we never know what we are talking about, nor whether what we are saying is true'. Courant and Robbins (1941) attempted to be rather more helpful in saying, 'Mathematics as an expression of the human mind reflects the active will, the contemplative reason, and the desire for aesthetic perfection. Its basic elements are logic and intuition, analysis and construction, generality and individuality.' Much earlier, probably the most eminent mathematics educator in his day, Nunn, wrote (1914), 'mathematics . . . [is] on the one hand a means by which man has constantly increased his understanding of his environment and his power of manipulating it, and on the other hand a body of pure ideas, slowly growing and consolidating into a noble rational structure', and thus, in relation to aims, 'our purpose in teaching mathematics in school should be to enable the pupil to realize . . . this two-fold significance of mathematical progress'. The title *Mathematics: Queen and Servant of Science* (Bell, 1952) immediately suggests this same dual purpose. The view of Sawyer (1955) that 'Mathematics is the classification and study of all possible patterns' seems to have led to many subsequent reiterations, perhaps with minor modifications like 'Recognition of pattern and . . . [its] subsequent communication . . . is the core of mathematics' (Biggs and Sutton, 1983). Mathematics certainly means many things to many people: an organized body of knowledge, an abstract system of ideas, a useful tool, a key to understanding the world, a way of thinking, a deductive system, an intellectual challenge, a language, the purest logic possible, an aesthetic experience, and a creation of the human mind being some of the many possible elements of a definition. For Davis and Hersh (1983) the definition of mathematics changes over time, for 'each thoughtful mathematician within a generation formulates a definition according to his lights'.

Perhaps the most fundamental way in which this debate continues today in relation to teaching school mathematics is the product–process debate. Mathematics is not only an organized body of knowledge (product), it is said, it is an often disorganized and untidy creative activity (process). A good elaboration on what might be considered to be the dimensions of the process aspects of mathematics can be found in Bell, Costello and Küchemann (1983). Debate on this dichotomy is often seen as a relatively recent phenomenon; many older (perhaps now referred to as traditional) views of mathematics would have been unlikely to have acknowledged that there was a place in school mathematics for engaging in the activities of mathematics, that is, acting like a mathematician, in quite the same way as is widely, though not universally, accepted today. Yet, over a considerable period of time, there have been statements from learned bodies which basically reiterate a kind of process view, for example that 'a larger place . . . must be found for those activities which we believe opinion would generally agree to call creative', and that 'subjects should be pursued actively' (Spens, 1938). There is a connection here with contemporary views on effective learning, and this issue is developed in Chapter 3. Modern reactionary educational views, variously described as traditional or 'back to basics', would be likely to view the process side of mathematics with suspicion. Thus, it certainly seems likely that, whatever view one holds about the nature of mathematics and how it should be reflected in what we teach in school, it will have some effect on one's own particular aims of teaching mathematics.

Another contemporary aspect of the discussion of the nature of mathematics is concerned with the impact of modern technology. The contribution which calculators can make to the learning of mathematics has already been referred to, but the issue runs deeper. We now have technology, becoming more accessible all the time, which will do a great deal of the mathematics that teachers have previously laboured to teach to generations of pupils and students. What we thought of as essential mathematics can now be done by machines, so was it ever really mathematics? Computer algebra systems (see Chapter 12) can be viewed as making it unnecessary for us to teach algebra, trigonometry and calculus in the way we have been doing up to now. So what is the essential nature of mathematics nowadays, if it is not learning how to do long divisions, quadratic equations, partial fractions, integration and the like? Has technology changed what we should view the nature of mathematics to be?

A CLASSIFICATION OF AIMS FOR MATHEMATICS TEACHING

In a powerful and what many would regard as classic statement of 'facts to be kept in mind in the teaching of mathematics for citizenship', Smith (1928) elaborated seven reasons for teaching mathematics. In essence these were that:

- every educated person should know what mathematics means to society and to our race, what its greatest uses are;
- it has high value as a mental discipline;
- it has intrinsic interest and value of its own – it has its own beauty and magic;
- it possesses truth which, in an ever changing world, is eternal and enduring;
- it enables us to understand our place in a world which contains such contrasts as the infinite and the infinitesimal;
- it came into being through the yearning to solve the mysteries of the universe and still works for us in that way;
- the history of mathematics is the history of the human race.

In comparison with what often seems an impassioned essay by Smith, all other statements of aims seem somewhat mundane. In the influential 1901 declaration by Perry (see Ministry of Education, 1958), for example, the entire emphasis appears at first sight to be on usefulness. This is in sharp contrast to Smith who includes reference to usefulness only within the broader category of awareness of the value of the subject. The Cockcroft Report (Cockcroft, 1982) places great emphasis on usefulness in the first two parts of the statement that the mathematics teacher has the task:

teaching

- of enabling each pupil to develop . . . the mathematical skills and understanding required for adult life, for employment and for further study and training . . .;
- of providing each pupil with such mathematics as may be needed for his study of other subjects;
- of helping each pupil to develop . . . appreciation and enjoyment of mathematics itself and . . . of the role which it has played and will continue to play both in the development of science and technology and of our civilization;
- above all, of making each pupil aware that mathematics provides him with a powerful means of communication.

The NCTM (1989) differentiates between 'societal goals' and 'goals for students', thus again endorsing the division into personal goals and the goals of society. The new social goals from the NCTM include mathematically literate workers, lifelong learning, opportunity for all, and an informed electorate, and the goals for students are that they learn to value mathematics, that they become confident in their ability to do mathematics, that they become mathematical problem solvers, that they learn to communicate mathematically, and that they learn to reason mathematically. Scopes (1973) defined four categories of aims, namely utilitarian goals, social goals, cultural goals and personal goals. Many other writers, for example Wain (1989), have given their own variation on the main categories of reasons for teaching mathematics. It is perhaps of most value to attempt to look at a number of the commonalities from among these. The five selected for consideration are listed below:

- mathematics is useful;
- mathematics is important in our lives and its place needs to be understood;
- mathematics trains the mind;
- mathematics is a powerful means of communication;
- mathematics is enjoyable and has aesthetic value.

UTILITARIAN AIMS

Perry, referred to above (Ministry of Education, 1958), maintained that the study of mathematics began because it was useful, continues because it is useful, and is valuable to the world because of the usefulness of its results. This must not be taken too literally, for Perry's uses included 'giving mental pleasure' and 'teaching . . . the importance of thinking things out', although the overall impression is that he placed considerable emphasis on the utilitarian aspect, in all its forms. Certainly, once education was becoming available to all, the utilitarian view appeared to hold sway for the 'common pupils', according to Maclure (1986). The mathematical aims of the elementary schools were intended to be based on the practical arithmetic skills needed in everyday life, on the wants of the working man (women, it was believed, did not really need much mathematics at all). In due course, however, the value of so much arithmetic began to be questioned, and we are still questioning it today. The Hadow Report (Hadow, 1926) called for a replacement of much traditional arithmetic because 'the amount of . . . indispensable arithmetical knowledge . . . is in reality comparatively small and would not in itself justify the time given to the subject'. Spens (1938) included many statements which related to the supposed usefulness of mathematics, and which were critical of an approach to mathematics teaching which was governed solely by utilitarian considerations. Dienes (1960) also commented on this issue in saying, 'If the requirements of everyday life determined the contents of our mathematics syllabuses there would surely be little mathematics in them.' The broad aim of equipping pupils for life after school would, I think, be accepted by all. But this does not imply that, having accepted this aim, there will be complete agreement about the detail and amount which should be taught under this heading. The NCTM (1989) suggested that we need 'mathematically literate workers', and quotes the definition of mathematical expectations of new employees provided by Pollak:

- the ability to set up problems with the appropriate operations;
- knowledge of a variety of techniques to approach and work on problems;
- understanding of the underlying mathematical features of a problem;
- the ability to work with others on problems;
- the ability to see the applicability of mathematical ideas to common and complex problems;
- preparation of open problem situations, since most real problems are not well formulated;
- belief in the utility and value of mathematics.

This is the foundation for an ambitious set of aims, the implementation of which would require devoting time to much more than what is strictly utilitarian.

There are, of course, other faces of usefulness, which Scopes (1973) summed up as (1) foundations for subsequent more advanced study of mathematics, and (2) tools for other subjects. To some extent, new mathematics is inevitably built on mathematics learned previously, thus usefulness cannot be denied as an internal requirement, though the detail of any syllabus needs to be kept under constant review, for needs do change. However, the needs of subsequent mathematics do not present a reason for teaching mathematics in the first place. And there is also something of a question mark against the other use, for it is possible to question whether one of the aims of teaching mathematics really is to service other subjects. Why should mathematics service the requirements of, say, physics? Does physics service any requirements of mathematics? Why should not mathematics be taught solely for the sake of developing its own internal structure and logic, or even for providing enjoyment? Why cannot physics provide for its own needs? If one accepts both divisions of the 'definition' of mathematics provided by Nunn (quoted earlier) and subsequently endorsed by others, then one might believe that the service aspect of mathematics was legitimized. All knowledge is, after all, connected, and subject divisions are to some extent artificial, so it can be argued that it is important to provide links between subjects. However, a link is not the same as a service, and not everyone agrees that mathematics has any obligations outside of itself. Thus we are reminded once again of how aims, beliefs and even values are inextricably intertwined. The issue of mathematics and other subjects is taken up in Chapter 11.

THE IMPORTANCE OF MATHEMATICS IN THE WORLD

The indisputable fact is that mathematics is vital to the maintenance of satisfactory living standards. It is mathematics which underpins the science and technology that support modern society. It would seem to be a legitimate aim for educators to wish that pupils will come to an understanding of how society works, and this implies an understanding of how mathematics provides support. We would perhaps wish to help pupils to learn that people have constantly sought to increase their understanding of their environment and their ability to manipulate and control it. An aim from Zambia is 'to assist the child to understand mathematical concepts in order . . . to . . . better understand . . . the environment'. The Mathematical Association (1919) wrote:

> in so far as mathematics is concerned, . . . education should enable [a child] . . . not only to apply . . . mathematics to practical affairs, but also to have some appreciation of those greater problems of the world, the solution of which depends on mathematics and science.

Similarly, from Spens (1938) we have, 'without some acquaintance [with mathematical thought] . . . much that is fundamental in modern life is unintelligible'.

The problem for teachers is precisely what and how much of the available knowledge and methods of mathematics is appropriate in order to provide enlightenment. It may be that very little is needed in the way of content, but some will be. Smith (1928) included reference to this problem and also pointed to the need for pupils to know enough about mathematics in order to make an informed decision about continued study of the subject:

A subject even so essential as [mathematics] in our world economy today need not be mastered by every citizen . . . [but] every educated man or woman should know what mathematics means, what its greatest uses are, and something of its soul, and should thus be able to decide whether or not he or she cares to pursue its study beyond the point of acquiring this elementary knowledge . . . everyone should know . . . what mathematics means, at least for the reason that the world uses it so extensively.

Spens (1938) said, in similar vein, 'we cannot leave it to the pupil to determine whether he is to elect for mathematics . . . until he knows what the subject means.'

Spens also included a number of references to the issues relating to balance between mathematics for utility and mathematics for awareness, for example the following:

As taught in the past, [mathematics] has been informed too little by general ideas, and instead of giving broad views has concentrated too much upon the kinds of methods and problems that have sometimes been stigmatised as 'low cunning'.

and also:

[A mathematics course should] . . . not aim at completeness in a limited sphere, it will not pursue mastery of technique beyond what is required for use, but [should] endeavour in its choice of topics and ideas to show how man has faced and solved problems the solution of which was vital to his progress, and [should] take care to introduce the pupil to the tools which are most useful for the problems of today,

and:

We deem it unfortunate that any pupil should leave . . . school without some inkling of the stupendous influences which ideas of abstract thinkers have had upon the world, and without some notion of the aims and techniques of exact thought.

Understanding, modelling and to some extent manipulating the world would be an aim of education accepted by many. This would certainly seem to add weight to the justification for including mathematics within the curriculum. It is, however, open to question whether what goes on in classrooms around the world does justice to this aim. Could we do better?

MATHEMATICS TRAINS THE MIND

For many years mathematics has suffered from a belief that 'when [it] is not directly useful, it has indirect utility in strengthening the powers of reasoning or in inducing a general accuracy of mind' (Spens, 1938). According to Watson (1913) Vives described mathematics as a subject to 'display the sharpness of the mind'. Isaac Watts wrote that 'If we pursue mathematical speculations, they will inure us to attend closely to any subject, to seek and gain clear ideas, to distinguish truth from falsehood, to judge justly, and to argue strongly' (Howson, 1982). Perry's 1901 address included, as obvious forms of usefulness, 'in brain development', and 'in producing logical ways of thinking' (Ministry of Education, 1958). How valid are claims that mathematics provides a kind of mental training which justifies its place within the school curriculum even when other uses cannot be found? This is a difficult question. It is now generally accepted that the literal claim is much too simplistic, that if there is benefit it is more subtle than was often recognized in the past, and that blind belief in the value of mathematics as a mental discipline is dangerous. Godfrey, a prominent mathematics educator of his day, wrote in Godfrey and Siddons (1931):

> when it is said that mathematics develops the memory, the logical and reasoning faculty, the power of generalisation, develops all these powers as applied not only to mathematics but also general activities – well, I hope that it may all be true, but I have not met with a proof.

The belief in mathematics (or Latin) as useful mental discipline with wide value across other subjects is associated with 'faculty psychology', which has long since been repudiated by educational psychology. But at the same time, it is reckoned that some kinds of transfer of training can and do take place in learning. Without any kind of transfer, children could only be expected to practise what we actually teach, and, 'The span of their learning could never exceed the range of situations or problems actually encountered in the course of instruction' (Shulman, 1970). With the sheer quantity of knowledge now available to learners, this would seem to suggest that there would not be sufficient time within a lifetime to learn enough. Thus it is now accepted that transfer can take place under particular circumstances, and what disagreements remain revolve around such questions as breadth of transfer, and whether it is only specific products and narrow processes which can be transferred or whether it extends to the generalization of broad principles, general strategies of inquiry, motivations and attitudes. One useful distinction is between vertical transfer, for example within a subject, and lateral transfer, for example across subjects. The learning theory of Robert Gagné (1970) depends on vertical transfer, but Bruner (see Shulman, 1970) has expressed his belief in much more massive transfer of broad principles and strategies from one domain or topic to another. This is not the same as faculty psychology, which has in the past been used to claim that the study of geometry makes one a better logical thinker *per se*. Bruner's theory is that broad transfer of training occurs when one can identify in the structures of the subject matters basic, fundamentally simple concepts, principles or strategies, which, if learned well, can be transferred both to other topics within that discipline and to other disciplines as well, for example the concepts of conservation and balance.

Thus the issue of transfer is not completely resolved, but this is not an excuse to view mathematics as a way of disciplining the mind, and respected opinion from within mathematical education agrees with the consensus from psychology. Godfrey and Siddons (1931) included the statement, 'In things of the mind, as in things of the body, we are still fairly ignorant of the effect of any particular course or diet; till we know a good deal more, our safest guide will be appetite.' More recent opinion concurs. The justification for teaching mathematics because it trains the mind is open to more doubt than any other reason. However, in another sense it is legitimate to speak of training as 'mental discipline', for 'it involves the submission of the pupil to the influences of the great tradition' and 'the faithful study of . . . major subjects does . . . impart some virtue to the mind' (Spens, 1938).

THE IMPORTANCE OF MATHEMATICS AS A LANGUAGE

A language is a means of communication. Mathematics has increasingly become a vital means by which ideas are conveyed, and thus some would claim that mathematics is a language. Many people would wish to claim that this dimension of mathematics is one of the most important, or even that the single most important reason for teaching mathematics is because it is a language. HMI (1985a) wrote, 'The main reason for teaching mathematics is its importance in the analysis and communication of information and ideas.' Cockcroft (1982) was equally clear that mathematics was an important means of communication (see earlier quotation). A point often made is that mathematics is a unique universal language which transcends social, cultural and linguistic barriers, having symbols and syntax that are accepted the world over. It is even claimed that extraterrestrial communication would have to take place via mathematics, for 'the most hopeful symbol we could give to attract the attention of a world much older than our own, and probably more refined, would be the figure of the theorem of Pythagoras' (Smith, 1928).

One of the main reasons for emphasizing to all pupils the importance of mathematics as a means of communication is its use in this way in the media. Newspapers and television, the main channels through which information is conveyed to most people, make great use of graphs and tables, the former sometimes potentially misleadingly presented. Most children are capable of understanding the elementary graphical forms when they often have great difficulty with the symbolism of algebra. It is, however, the concise symbolism which is another feature of mathematics as a language, for 'it has been stressed that mathematics provides a means of communicating which is concise and powerful, not least because of the use of symbolism' (Mutunga and Breakell, 1987). From Marjoram (1974) we have:

> mathematics is a subject in which terms are carefully and unambiguously defined and in which language must be used with care, . . . knowledge is accreting at a formidable rate . . . [and] words alone are becoming too extravagant, slow, cumbrous and inadequate [as] a means of conveying new ideas; verbiage is imploding into symbolism.

Thus it is claimed that we have a duty to teach mathematics because of its value as a concise and internationally understood way of communicating certain kinds of ideas which cannot sensibly be communicated through other means. The wider claim that mathematics is a language might be open to more dispute, however.

THE IMPORTANCE OF DEEPER APPRECIATION

Many mathematics teachers believe that mathematics can provide pleasure and enjoyment. Thus, an aim in teaching mathematics could be that the learner comes to enjoy mathematics, gaining pleasure from the scope and intricacies of the subject, from its patterns, and from what it can reveal about learners and their worlds. In the extreme it can be claimed that mathematics can provide a significant input into aesthetic appreciation. Many believe that there is as much about mathematics which is aesthetically pleasing as there is about any other subject which might more frequently come to mind when aesthetics is under discussion. Sadly, most pupils, and indeed many mathematics undergraduates, do not see it that way, so we cannot claim great success here. If aesthetics is an aim of education, however, mathematics certainly has as much right as any subject to claim space, and it is then up to educators to try to achieve this aim. Perhaps too much effort is extended on the utilitarian and not enough on an appreciation of the wonders of the subject. Spens (1938) wrote, 'no cultural tradition is adequately represented by teaching which fails to give a proper place to . . . the sense of wonder or romance.' Scopes (1973) wrote of the aesthetic fascination of mathematics through the contemplation of shapes and patterns, the elegance of proof, the effective use of symbols and the unity of seemingly different branches of mathematics. The SMILE scheme (1986) included as an aim, 'to appreciate mathematics as a creative, aesthetic, intellectual activity'. Smith (1928) wrote, 'let us see . . . that the poetic side of mathematics is recognized as well as the practical side: let us . . . show the world how to use its leisure as well as how to turn the restless wheels of industry' and gave as examples the beauty of symmetry, the relationship between natural form and Fibonacci, the structure of snow crystals and 'the poetry of the complex sixth roots of unity'. We should lead the student to those places 'because they rouse his soul to the truths which endure', he claimed. In contemporary mathematics the Mandelbrot and Julia sets generated within the study of chaos, for example, remind us once again of the aesthetic appeal of mathematics. Chaos, as a relatively new branch of the subject, is discussed in Chapter 6.

A further aspect of a deeper appreciation of life and of our world is through history and, 'the history of mathematics is the history of the human race' (Smith, 1928). From the same author we have a claim relating to history which falls within the domain of aesthetics:

> In the history of the world, mathematics had its genesis in the yearning of the human soul to solve the mystery of the universe in which it is a mere atom . . . [and] . . . in the minds of those who followed the courses of the stars, . . . and [it seems] to have had its first real development in the effort to grasp the Infinite . . . even today it is the search into the Infinite that leads us on.

Spens (1938) said that mathematics is 'one of the main lines which the creative spirit of man has followed', and that 'When ideas are introduced in their historical setting it is possible to see how one idea grows out of other ideas in response to a realised want, and how ideas are fitted together and contribute to the gradual growth of a living structure.' Scopes (1973) said, 'Every student should be made aware of some of the major strands in the history of mathematics, and how this has influenced the thought processes of successive generations.' Among his illustrations of completely original thought processes which have revolutionized the world he included zero, coordinates, continuous change, computing, and the more inward-facing aspects like structure and axiomatics. Quality of life, leisure and aesthetics are all legitimate considerations when defining the aims of education. Many would say that mathematics is needed because of the particular unique contribution it can make to these dimensions.

REFERENCES

Alexander, R., Rose, J. and Woodhead, C. (1992) *Curriculum Organisation and Classroom Practice in Primary Schools*. London: DES.

Bell, A. W., Costello, J. and Küchemann, D. E. (1983) *Research on Learning and Teaching*. Windsor: NFER-Nelson.

Bell, E. T. (1952) *Mathematics: Queen and Servant of Science*. London: G. Bell & Sons Ltd.

Biggs, E. and Sutton, J. (1983) *Teaching Mathematics 5 to 9*. Maidenhead: McGraw-Hill.

Bruner, J. S. (1966) *Toward a Theory of Instruction*. Cambridge, MA: Harvard University Press.

Cockcroft, W. H. (1982) *Mathematics Counts*. London: HMSO.

Courant, R. and Robbins, H. (1941) *What Is Mathematics?* London: Oxford University Press.

Curriculum Development Centre, Malaysia (1975) *The Goals of Education in Malaysia*. A circular distributed to schools.

Davis, P. J. and Hersh, R. (1983) *The Mathematical Experience*. Harmondsworth: Penguin.

Dienes, Z. P. (1960) *Building Up Mathematics*. London: Hutchinson.

Gagné, R. M. (1970) *The Conditions of Learning*, 2nd Edition. New York: Holt, Rinehart & Winston.

Godfrey, C. and Siddons, A. (1931) *The Teaching of Elementary Mathematics*. Cambridge: Cambridge University Press.

Hadow, H. (1926) *The Education of the Adolescent*. London: HMSO.

HMI (1979) *Mathematics 5–11*. London: HMSO.

HMI (1985a) *Mathematics from 5 to 16*. London: HMSO.

HMI (1985b) *The Curriculum from 5 to 16*. London: HMSO.

Howson, A. G. (1982) *A History of Mathematics Education in England*. Cambridge: Cambridge University Press.

Layton, D. (1978) Curriculum Theory. In G. T. Wain (ed.), *Mathematical Education*. Wokingham: Van Nostrand Reinhold.

Maclure, J. S. (ed.) (1986) *Educational Documents England and Wales 1816 to the Present Day*. London: Methuen.

Mager, R. F. (1975) *Preparing Instructional Objectives* (issued 1962 as *Preparing Objectives for Programmed Instruction*). Belmont, CA: Fearon.

Marjoram, D. T. E. (1974) *Teaching Mathematics*. London: Heinemann.

Mathematical Association (1919) *The Teaching of Mathematics in Public and Secondary Schools*. London: G. Bell & Sons Ltd.

Ministry of Education (1958) *Teaching Mathematics in Secondary Schools*. London: HMSO.

Mutunga, P. and Breakell, J. (1987) *Mathematics Education*. Nairobi: Kenyatta University and London University.

NCTM (1989) *Curriculum and Evaluation Standards for School Mathematics*. Reston, VA: NCTM.

Nunn, T. P. (1914) *The Teaching of Algebra (Including Trigonometry)*. London: Longmans, Green and Co. Ltd.

Orton, A. (1992) *Learning Mathematics: Issues, Theory and Classroom Practice*, 2nd Edition. London: Cassell.

Russell, B. A. W. (1921) *Mysticism and Logic*. London: Allen and Unwin.

Sawyer, W. W. (1955) *Prelude to Mathematics*. Harmondsworth: Penguin.

Scopes, P. G. (1973) *Mathematics in Secondary Schools*. Cambridge: Cambridge University Press.

Sewell, B. (1981) *Use of Mathematics by Adults in Daily Life*. Leicester: ACACE.

Shulman, L. S. (1970) The psychology of learning mathematics. In E. G. Begle (ed.), *Mathematics Education*. Chicago: NSSE.

SMILE (1986) *GCSE Aims and Objectives*. London: SMILE.

Smith, D. E. (1928) Mathematics in the training for citizenship. In the NCTM Third Yearbook, *Selected Topics in Mathematics Education*. Washington, DC: NCTM.

Sosniak, L. A., Ethington, C. A. and Varelas, M. (1991) Teaching mathematics without a coherent point of view. *Journal of Curriculum Studies* **23**(2), 119–31.

Spens, W. (1938) *Report of the Consultative Committee on Secondary Education*. London: HMSO.

Wain, G. T. (1989) Mathematics. In P. Wiegand and M. Rayner (eds), *Curriculum Progress 5–16*. Lewes: Falmer Press.

Warnock, M. (1978) *Special Educational Needs: Report of the Committee of Enquiry into the Education of Handicapped Children and Young People*. London: HMSO.

Watson, F. (1913) *Vives on Education*. Cambridge: Cambridge University Press.

Watson, J. M. (1992) Long division: a sense of déjà vu. *Mathematics in School* **21**(3), 42–3.

Whitehead, A. N. (1932) *The Aims of Education and Other Essays*. London: Williams and Norgate Limited.

Chapter 2

Mathematics Education and Society

Geoffrey Wain

INTRODUCTION

There can be no doubt that formal education in general is rooted in a social and cultural context. From the moment that universal education is established in any country the provision of education becomes a political matter subject to government decisions about policy and financing. Governments see the purposes of education as concerned with broad social issues such as nation-building or the provision of an educated population which will enable the country and its economy to develop and thrive. They are therefore keen to ensure that educational investment is appropriate to its purpose and that is usually brought about by prescribing the curriculum to be followed. Such curricula inevitably see pupils as a homogeneous group to be processed together in order to bring about certain outcomes that can then be used as the measures of success or otherwise. But education is notoriously difficult to assess in terms of value for money and is extremely expensive. In the end the outcomes of the process are individuals equipped with a diverse range of skills, knowledge and attitudes, not groups of individuals with identical characteristics. It is also unfortunate that education has often not appeared to fulfil the expectations of governments in bringing about modernization or industrialization and there would seem to be little doubt now that it is far easier to build a school system than to modernize society (Simmons, 1980). The idea that education alone can change a society would seem now to be thoroughly discredited. Obtaining qualifications that lead to unemployment or that have no immediate use in society can lead to great disillusionment. Priorities, therefore, between educational and other investment need to be carefully considered and the interrelationship between education and the true needs of society carefully analysed.

Pupils and their parents view education differently from governments. Education to them may be seen in a narrow context as contributing directly to their needs within a limited part of society or, more widely, to providing certificates that give access to new possibilities and increased wealth. Their viewpoint will depend crucially on their position in society and the extent to which their own aspirations are congruent with the broad aims that have determined the curriculum. From their perspective education will be evaluated by the extent to which it enables them to survive and thrive in society. At worst they may see education as irrelevant to their needs.

It is, of course, likely that the socio-political needs of society and the needs of some individuals will be broadly in agreement. The extent to which this is the case will depend upon the stage of development and the complexity of the society, the home background of the pupils, the opportunities for work or further education and the extent to which the educational system really does, in practice, meet the needs of both the society and the individuals within it. The extent to which, in so many cases it does not meet these needs is evidenced by many examples, such as the disaffection of many school pupils in Western societies (Mellin-Olsen, 1987) and the failure of formal education to meet the needs of rural communities in many developing countries (Simmons, 1980).

Within education in general, mathematics holds a key position. Together with language and science it forms part of a core of school subjects that are accepted as compulsory world-wide and, as such, must be considered in relation to the general issues raised above. The purpose of this chapter is to explore some of those general issues in so far as they concern mathematics. It might, of course, be argued that the very nature of mathematics is asocial, the theorem of Pythagoras being true no matter what the social context, but when mathematics is part of general education it is immediately subject to political decisions about aims and purposes (see Chapter 1), and the nature of the mathematics curriculum is determined by considerations other than those of a purely mathematical nature. That this is so has been accepted fully in recent years (Bishop, 1988; Wilson, 1981). Of course, it is possible that a curriculum may be chosen which presents mathematics in an abstract way with no apparent links to social context, but making that choice is, in itself, a decision that would need to be justified in terms of the various needs involved, including those of the pupils and of society. Such a decision might be appropriate but must still be made in a socio-political context.

Many syllabuses have, of course, been constructed with the needs of abstract academic mathematics solely in mind. They have been determined primarily by the needs of the next stage of mathematical education; defined from the top as it were. The payoff of success for a few is intrinsic satisfaction and progression through the system, but for the majority the only success might be the knowledge that a particular certificate has been obtained by memorizing material that has no perceived value. Many may achieve no success at all and large numbers of pupils will leave formal education with the view that mathematics is a meaningless subject. The idea that syllabuses could be designed 'from the bottom up' is a very recent one (Cockcroft, 1982) and requires of the designers a much more careful consideration of general educational and social needs. Above all, it would seem reasonable that mathematics should have relevance and meaning to all learners at the time of learning and also relate in some way to the society in which they live and to the future needs of that society. This implies that the learner is being provided with an induction into mathematics in all the forms that are appropriate to society.

Mathematics is, and always has been, alive within society in many forms. The full extent to which it is so will be considered later, but clearly one of the challenges that faces those constructing a school curriculum is to make sure that the mathematics taught relates naturally to the reality of mathematics in all its aspects in society, including academic developments, uses in work and at home, as an important part of our culture and as an aspect of many everyday activities. Of course, both society and mathematics are continually changing, so that the link between them must continually be reassessed.

This is true in all societies whether industrialized or developing, although in the latter the situations that can arise may appear in very dramatic forms, as has been shown in the cases, for instance, of the Kpelle in Liberia (Gay and Cole, 1967), the Navajo in North America (Pinxten, *et al.*, 1983), and many others where the juxtaposition of academic mathematics with the ethnic mathematics of the society may suggest that no links between the two are conceivable. It would be a mistake to assume that similar problems, albeit arising in different circumstances, do not exist in developed societies (Mellin-Olsen, 1987). Many pupils in affluent societies acquire poor attitudes to the subject, which is seen as meaningless by many. This is compounded by the fact that inability at mathematics is often seen as socially acceptable where illiteracy would not be. Young people in economically developed societies now grow up in a culture deeply influenced by the presence of computers and other technology, and the implications for mathematics education are great but have only just begun to be considered. Meanwhile young people will inevitably see the mathematics of school as inappropriate if they are aware that machines have superseded the methods they are being taught. This point will be considered again later.

THE NATURE OF SOCIETY

One of the problems in considering the mathematical needs of society is that it is notoriously difficult to provide a definition of society that is helpful. If everyone in a country were involved in agriculture then it would seem reasonable to call that society agricultural and to make educational decisions based on that description. But such a simplistic description is hardly likely to be completely accurate in practice. For example, in England in the eighteenth century about 90 per cent of all workers were involved in some aspect of farming, so that it might seem reasonable to call the society of that time an agricultural one. That would clearly be misleading, however, since other groups of considerable importance were present in society, including a very strong academic scientific community based on old-established universities. Whatever the educational needs of the agricultural community, they would hardly fit easily alongside those of the other groups. The effective schooling available at the time was not aimed at the majority, who received little education. They learned from their parents, the church, their employers and their local community, and, in the main, what they learned consisted of traditional skills handed down within a stable social context. A great deal of what was learned was not 'taught' in the sense that would be used in most countries today where teaching is institutionalized in the schools. Such a situation still exists in many countries where a great deal of what may be considered useful learning takes place outside the school.

Towards the end of that century the demands for general education in England grew alongside the industrial revolution. The formal system of education that finally emerged in 1870 was essentially a response to changes in society. Those changes were not brought about by developments in education, although education fuelled the change and provided workers with some of the basic skills that were required for the new industries. A particularly significant point to note is that from this time onwards the educational process no longer relied solely on traditional transmission of skills through the family and the local community. In fact the formal educational process introduced the new possibility for many that education could lead to social mobility, and opened up a wider range of opportunities for individuals.

By the early twentieth century most workers in England were involved in industrial manufacture, so that now the society could perhaps be called industrial. Such a description would again be a gross oversimplification, more so than it would have been before, because one of the significant new developments that had occurred and which has continued right through the century has been the growth of what is usually referred to as the post-industrial society, typified by the existence of different sectors alongside one another. Thus there is now an agricultural sector employing about 3 per cent of the population, an industrial sector with about 15 per cent and declining, and the rest made up by a great diversity of work, much of it often called the service sector. Even in one sector, such as the industrial, the range of work is immense and increasingly demands of the educational system a flexibility of approach that enables future employees to cope with many very different work situations – a point that will considered later.

It would be wrong to suggest that England had succeeded in matching its education system to its developing society in anything but a crude way, even though the education system has been reasonably responsive to the changes that have occurred. The same can probably be said of any other country, although the success of the match will vary.

The point to make is that an education system must be designed in relation to the reality of society and must be sensitive to changing needs. The difficulty is in deciding what that reality is for mathematics and what the relationship should be. It is probably now more important than ever before to get the relationship right as quickly as possible, because the rate of change in many societies is so great that education is always in great danger of lagging very far behind. This latter point is well illustrated by the introduction of the electronic calculator in the early 1970s, which is now accepted by societies worldwide as the means of carrying out calculations in industry and commerce, and yet, twenty years on, many schools have not accepted its existence and many teachers of mathematics cannot use it. There are many other aspects of the relationship of mathematics education to society which need to be considered, and the rest of this chapter will be concerned with some of them.

MATHEMATICS IN SOCIETY

As has been mentioned earlier, mathematical education has frequently been viewed from 'above'. The needs of the academic world have taken precedence in determining what mathematics should be taught in schools. That this should be so illustrates the main problem facing the design of a mathematics curriculum; that is, how to reconcile the needs of those pupils who will go on to study the subject at a high level with those whose mathematical requirements are much more modest. The main purpose of this chapter is to review some of the ways that reconciliation can be brought about through a consideration of the real needs of society in terms of mathematics, and how satisfying those needs can bring benefit to all pupils, albeit in different ways.

Mathematics formalized within education has often ignored completely the mathematics that exists outside. Some examples will illustrate the point. In writing about the history of the theory of probability, Todhunter (1865) considered that the subject began with Fermat and Pascal. It is, of course, correct to say that the mathematical theory of probability began with them, but there is ample evidence that probability was alive and well in society long before they laid the mathematical foundations. Rabinovitch (1973) gives an interesting account of the use of probability as reported in ancient and medieval Jewish writing and demonstrates that arguments based on the addition, multiplication and significance of probabilities were employed by rabbis many centuries before Pascal and Fermat. One of the oldest activities, that of gambling, has also used ideas of chance and there is evidence that many gamblers understood the way that probabilities behaved long before the seventeenth century.

Another frequently quoted example is the case of Samuel Pepys who, after education at St Paul's School and Cambridge University, was not able to do simple arithmetic and, in 1662, was taught by Cooper, the mate of the ship called the *Royall Charles*. The interesting point here is that arithmetic was well known to many people (in this case a sailor), but not to someone like Pepys who had a mathematics degree from a university. Those who built the ships for the navy also knew their arithmetic, and it was to cope with the builders and their accounting that Pepys, as secretary to the navy, was motivated to study the subject. The picture that emerges is of academic mathematics, well recorded and documented, representing the tip of an iceberg of mathematical activity alive in society in general but much less well documented.

A similar situation existed in Victorian England, where the process of industrialization employed mathematics at work in a way not encountered before. The design, for example, of engines, pumps and lifting equipment, and the accounting and recording that were needed, all testify to the existence of many practitioners of mathematics at a variety of levels. It is beyond the scope of this chapter to detail further the historical relationship between mathematics in use and academic mathematics. Suffice it to say that general education has always chosen to concern itself with the academic aspects of the subject rather than what Griffiths (in Wain, 1978) has termed the artisan.

The examples given above have been taken from periods in English history. Similar accounts could be given of the situation elsewhere, but there is still a great deal of research to be done in exploring the mathematics of society in many countries as it exists at present and through history. Some useful references are d'Ambrosio (1985), Pinxten *et al.* (1983) and Bishop (1988).

The issues that are important in contemporary society are first to determine what mathematics exists and has significance for people in general, and then to determine the extent to which education can and should take account of it. There are several different aspects to take into consideration, and they will be dealt with in what follows.

MATHEMATICS IN DAILY LIFE

The mathematical needs of people in their daily lives will obviously vary depending upon what is meant by 'daily life'. For our purposes we will consider what is often referred to as survival mathematics; that is, the mathematics that is required to make sense of normal, everyday situations. The definition is still elusive and will depend completely on the normal activities of the society. Thus, for example, to get by in an industrialized society will need such skills as the ability to recognize numbers in a variety of contexts, to use a calculator with understanding, to tell the time and read timetables, to understand and be able to work with the normal measures of weight, length, area and capacity, to understand the properties of simple geometric figures in practical settings, and to be able to make sense of data presented in a wide variety of forms. These skills enable a person to carry out the basic tasks of living, including being aware of time, being able to travel, to cook, to make sense of instructions to use a range of gadgets such as petrol pumps, cookers and so on, and also to be able to communicate ideas that require to be put in a simple mathematical form. Considerations of this type led to the construction of foundation lists of skills (Cockcroft, 1982).

The skills that are basic to survival will vary from society to society. The range that exists has been documented to some extent by the writers on ethno-mathematics. The implications for education are, of course, considerable. It is surely important that such skills are not ignored in an educational process or, far worse, alienated by the imposition of techniques that do not work for the user in practice (Sewell, 1981; Gay and Cole, 1967). An interesting approach that deals sympathetically with this issue is in Bishop (1988).

MATHEMATICS AT WORK

One of the obvious functions of education is to provide a preparation for the world of work. A detailed discussion of the form that preparation should take is beyond the scope of this chapter, since it would involve matters of general policy about the structure of school systems and levels of investment in training after school. What will be considered here are some of the issues about mathematics education that contribute to the discussion.

The form of the mathematics curriculum may, nevertheless, reflect matters of general policy. In particular it needs to be agreed what the aims of mathematical education are (see Chapter 1). It would be possible, for instance, to see mathematical education as providing skills that were of specific relevance to the world of work or, alternatively, as providing competencies that were of general applicability. Sadly, much of mathematical education does not seem to have succeeded in achieving either of these objectives, but, nevertheless, it has often been used as a general achievement measure for selecting people, notwithstanding the details of the education that they have received. Whatever the aims may be, it would seem reasonable to inquire what mathematics is used in the world of work and to use the outcomes of that inquiry to inform the curriculum development process.

There have been many reports of this type of inquiry, many of which occurred in the late 1970s in the face of criticism that schools were not producing people adequately trained for employment. The criticism occurred in many countries and fuelled the subsequent 'back to basics' movement, which is with us still. Many of the reports have shown that the link between school mathematics and the needs of employment is often very tenuous, and not conducive to the simple analysis that was often implied by the detractors. For many employees, for instance, the mathematical techniques that they use are not those that they have been taught in school. This has been particularly well documented by Fitzgerald (1981), who carried out a detailed analysis of the mathematics used in many occupations on behalf of the committee of inquiry set up by the British government in 1978 (Cockcroft, 1982). The terms of reference of the committee specifically asked for recommendations about the impact on mathematical education of the needs of employment.

Fitzgerald made the important point that many workers found it hard to recognize that they were using mathematics at all. An example will illustrate the point. A carpenter faced with the task of cutting $13\frac{1}{2}$-cm lengths from a 50-cm piece of wood will quickly realize that three can be cut and about 10 cm will remain. In school it is likely that the same problem would be expected to lead to the division:

$$50 \div 13\frac{1}{2} = 50 \times \frac{2}{27} = \frac{100}{27} = 3\frac{19}{27}$$

with the interpretation that 3 can be cut with $\frac{19}{27}$ left over! Even in such an elementary context the mathematics of the real world seems remote from its academic, formalized equivalent. The survey showed that skills used in the workplace were often seen as common sense once they were well understood, even though some were quite advanced in a mathematical sense. This illustrates how difficult it is to find out what mathematics an employee is actually using without observing the work process itself. Even then a particular task may be described inappropriately. For instance, a task that appears to be concerned with certain arithmetical and trigonometrical procedures might on closer observation be seen to depend critically on the worker being able to interpret geometrical situations that give rise to the calculations. These geometrical requirements might be very sophisticated. There are obviously dangers, therefore, in trying to interpret work needs without careful observation of the whole of a process. Even if that is done, as was the case in the Fitzgerald survey, there remains the danger that what is observed is assumed to describe the situation long after fundamental changes in working practices have taken place. Thus in 1985 Fitzgerald produced a further report designed to look specifically at the impact of new technology, which revealed that a remarkable change had taken place since the survey four years before.

In the first report Fitzgerald studied the mathematics used by school leavers in employment between the ages of 16 and 18 in a wide cross-section of employment situations. It is at this stage that the mathematics of school should be most significant since the skills acquired at school would seem to be the only ones immediately available for use. As time goes by the needs will change and the ability of the employee to adapt will depend less on what has been learned at school and more on experience and further education. It becomes increasingly difficult, therefore, to talk usefully about the interrelationship between school mathematics and that in employment as a person's career develops. Even so, for the new school leaver it was found that the mathematics used was rarely dependent on what had been learned in school. The employees had frequently devised their own idiosyncratic methods for getting answers to problems even though they had been taught specific algorithms at school which would have obtained the results. The idiosyncratic methods were sometimes devised by the workers themselves so that the process involved was clearer to them; at other times the methods had been taught by a senior person as part of the accepted tools of the trade. The way that the methods worked was not analysed to produce more efficient methods; once a method was seen to work it was employed in a purely instrumental way. Such methods are clearly context-related and unlikely to transfer to other situations. In many situations a more mathematical approach may not be appropriate or helpful, as can be seen above in the example on cutting wood. Work that was apparently mathematical was therefore reduced to routine activity that required no mathematical thinking as such and often depended heavily on the use of various aids. Fitzgerald makes the important point that to describe what is observed as mathematics at all may well be misleading, since no real mathematical activity or problem solving is involved. The implicit mathematics has thus been buried under procedures that remove the need for explicit mathematical activity. Fitzgerald therefore questions the value of some of the further mathematical training that is given to young workers when it seeks to explain mathematical processes in a theoretical way and when the reality of the workplace has no need for such explanations. Indeed, the trend described in the second report in 1985 towards the greater use of calculators and computers would suggest a further reduction in the need to understand the mathematics, and thus a further mathematical de-skilling of the people involved.

When the Cockcroft Committee of Enquiry reported in 1982 it drew heavily on the work of Fitzgerald and made a number of proposals about the consequences for teachers and schools. The findings were thought to be of great interest to teachers in helping them to understand the needs of the workplace, and it was recommended that teachers should liaise closely with employment in order to familiarize themselves with the work situation, presumably in order to bring greater relevance to their teaching. How such liaison might be established and carried through was given in great detail in Bird and Hiscox (1981). It was suggested that one of the outcomes might be to use teaching materials that would enhance the teaching situation and bring practical contexts to school work. It was also noted, however, that the materials produced by one teacher were unlikely to be useful to others who had not had the opportunity to see the work situation for themselves. The need for teachers to gain these experiences is now enshrined in various policies of government and teacher education criteria. As we have seen, however, the links are not at all obvious and there is always the danger of imposing mathematics inappropriately on practical situations. Getting the best balance requires a very sensitive approach to the design of curriculum materials. In addition, as witnessed by the Fitzgerald reports, the rate of change in the world of work is dramatic and any experience a teacher obtains may date very rapidly.

Ten years after the first report there is still no real evidence that using workplace related materials really does assist mathematical education or the transition into work of the school leaver. Teaching within too narrow a context has the danger of relegating mathematics to a subsidiary position, within which it may not be identifiable as mathematics at all and hence not lead to generalizable learning of the subject. More research is clearly needed to resolve these issues and to convince teachers of the value of placing mathematics in work-related contexts (Cleeves, 1991).

Mention has already been made of the snap-shot nature of each of the two reports by Fitzgerald. He detailed in the second the remarkable changes that had taken place in a period of only four years. Increased use of calculators and computers had changed the mathematical needs of the user. Most calculations were done on calculators, thus removing the need for written algorithms, tables of sines, etc., as these were contained within the machine; algebraic processes were dealt with in computer programs; control and measurement were often automated; and computer-aided work had removed the need for drawing and graphing skills. The residual mathematical skills were either very elementary, like the counting of money, or high level, as required in the design of programs and machines. Newly important were interpretative and estimating skills, keyboard skills, flexible skills in understanding geometrical ideas in two and three dimensions, and knowledge of iterative methods. The fact that there were such notable developments between the two reports suggests that there is now need for further research to prevent the findings of 1985 from being accepted as still true today when they may be becoming rapidly out of date.

The trend reported between 1981 and 1985 suggests a reduction in the need for routine skills and an increase in the ability to use new technology and the associated process skills such as interpretation, selection of key ideas, pattern spotting and adaptability. The particular case of the role of computers will be returned to later. Above all, perhaps, is the need to be able to respond to changes in working practices and to adapt to new demands, which suggests that there is a need for education that develops generalizable skills, critical awareness and independence in working, all of which have had a place in the creation of mathematics if not in its teaching. These are important issues to be taken into account in the classroom, and suggest that the use of more creative, investigational, practical and group work may develop the types of skills that now seem to be necessary. The teacher who experiences work attachment will probably learn that change is the most significant aspect of the world of work and that the mathematics observed in the workplace has only limited currency, since it is likely to be outdated as soon as it is enshrined in teaching materials.

Rapidity of change in the activity of the work place is such an important feature of the present situation that it needs to be considered a little further. The essence of many developing cultures, and of the culture in Britain in, say, the eighteenth century, was that change was either very slow or non-existent and most human activity depended much more than now on traditional methods handed down from generation to generation. In Fitzgerald's first study (1981), such a dependence on traditional methods passed on from experienced worker to novice was still evident in modern industry. Change as a highly significant part of culture is new and it may well be that the existence of change in itself is the catalyst that brings about a reconsideration of traditional approaches.

MATHEMATICS IN CULTURE

In the mathematical education literature in recent years, there has been considerable concern for the role of mathematics in general culture. The term 'ethno-mathematics' has been coined to describe that mathematics which is locked into a culture and which is used to accomplish many ordinary, everyday tasks. The tendency seems to have been to identify ethno-mathematics in developing societies rather than in industrial ones, although it will be shown later that this distinction is really unnecessary and possibly misleading. The literature (Gerdes, 1988; d'Ambrosio, 1985) contrasts the natural, folk, indigenous, non-standard mathematics of society with that taught in schools, thus showing how the needs of the academic curriculum push aside natural abilities to solve problems and handle mathematical ideas. But this is not just a problem for the developing world, as has been mentioned earlier. All societies exhibit mathematical activity in some form and Bishop (1988) has classified this under six headings: counting, locating, measuring, designing, playing and explaining. He illustrates how such a classification might be used in the process of designing a mathematics curriculum for any country, taking into account the obvious differences between societies; differences that are often revealed by the very ways in which these six activities manifest themselves. The intention of such a curriculum is described by Bishop as bringing about a process of mathematical enculturation, a bringing together in close union of the educational process and the whole of the culture within which it exists.

The intention is admirable. The concern is to unlock the mathematics 'frozen' in society (Gerdes, 1988) by careful anthropological analysis and then build bridges to it from the mathematics of the academic world, in order to produce a meaningful educational process or enculturation, using Bishop's term. There are problems facing such a programme, however, as has been mentioned in discussing the mathematics of the workplace. Traditional techniques used in context, although mathematical in nature to the experienced eye, may not be seen as such by the user, who may well see no place for mathematical analysis, particularly if it is thought to serve the purposes of the mathematical educator rather than the pupil. The instrumental skills of the quilt-maker who is building patterns by complex rearrangements of shapes are not likely to be enhanced by a course in transformation geometry, unless there is the most sympathetic relationship between teacher and taught and a very full knowledge of the skills of the quilt-maker on the part of the mathematician. The same dangers occur in any application of mathematics. The field of application may be sullied by having mathematics imported into it (Archenhold *et al.*, 1981) unless the person applying the subject has an adequate scientific model of the field of application (Crighton, in Orton, 1985).

The process of contextualizing the mathematical curriculum, then, has enormous potential for making the subject relevant, but there are dangers. As has already been mentioned, teachers may be capable of dealing with the mathematics involved but may find it difficult to deal with the context itself. Many examples used in mathematics in the past got round this difficulty by using what Pollak (1979) has called whimsical problems, examples of which are the famous problem on the rate at which taps fill a bath while the plug is out, and the one about the rate at which a number of men can dig a ditch compared with a different number of men. In such problems no attempt was made to use the situation in a realistic way – there was a conspiracy, to which all teachers of mathematics seemed to subscribe, that gave a localized social acceptability to meaningless activity of this whimsical kind. The problem of bringing the context more into the foreground of mathematical activity is, that new dimensions are added to the mathematical classroom that may be hard for the teacher to handle, or that convey messages that are irrelevant to the mathematics itself. Getting the balance between whimsical meaninglessness and overcontextualization that clouds the mathematical content is clearly one of the most important issues facing the teacher and the curriculum designer. The essential aim must surely be that a mathematics lesson has meaning to the learner at the time the lesson is in progress as well as for some future, perhaps distant, aim. This meaning may be achieved through a careful integration of mathematics with other subjects, or through the teacher being able to bring to the classroom adequate knowledge of other fields to enable the mathematics to be contextualized in a non-trivial way.

That mathematics education has singularly failed so many people in not providing meaning for the subject at all is a most sad commentary on the efforts of the teachers. The result has been that it is difficult to imagine how a subject could have achieved for itself such an appalling image as it now has in the popular mind.

THE POPULAR IMAGE OF MATHEMATICS

It is not necessary to spell out the evidence that leads one to say that the public image of mathematics is a poor one. The tragedy is that, after having spent between eleven and seventeen years in full-time education, many intelligent people are still able to claim that mathematics has always been, and will always be, a meaningless activity. This is in spite of the fact that in eleven years of full-time education a person will have received something over 1500 hours of mathematical teaching. To think that all that effort has led to a situation of fear and loathing is depressing. It is, of course, possible that the situation is not as bad as it seems, but that it is fashionable to say that mathematics was never understood. Whatever the truth of the matter, there is no doubt that the public image of mathematics could be better. One of the problems is that the world of the professional mathematician has not been very successful or willing in communicating to the public at large anything about the nature of mathematics. There would seem to be a professional disinclination to attempt to popularize the subject, although it is important to acknowledge the work of the few who have produced some excellent popularizing books. Even less has been done to bring to general notice the non-academic mathematics alive in society at various times.

The truth of the matter is that there are few ways in which the general public can get access to things mathematical. In the region of England around the place where this chapter is now being written, it is possible to get access to museums dealing with rail transport, industrial processes and industrial heritage, colour, photography, cinema, television and film, folk history, military history, coal-mining, farming, life during the Second World War, the Bronte family, and science, and to information about many other societies and cultures. There are concert halls and theatres, art galleries and sculpture exhibitions. In fact, it is possible to make visits to gain access in a lively way to aspects of every school subject – except, that is, for mathematics. Although, as has been pointed out above, the subject is embedded in all societies as well as being an important academic subject with a vast and growing literature, there has been little attempt to bring the subject alive in a popular way. It must be a matter of some concern that there is so little chance for people in general to gain access to information about a subject which is crucial to all societies. Of even greater concern is that teachers do not have ready access to the world of mathematics in easily digestible form, so that they can reflect in their teaching the living reality of the subject. It is, of course, acknowledged that there are a few museums and exhibitions that have been set up to do just what is suggested, but they are few and there is clearly room for many more.

It would be easy to claim that it is not possible to popularize the subject and that it is too difficult to make mathematical ideas accessible to the general public. This may be true if the intention is to convey information about, say, the latest theorems in differential geometry, but that is not what is intended here. The existence of mathematics in a wide variety of forms in society has already been discussed, and it is that wide variety that could be made more accessible to people. In a recent attempt to illustrate what might be possible, an exhibition called the PopMaths Roadshow toured the United Kingdom between 1989 and 1991 (Wain, 1992). Many thousands of people were attracted to it. Apart from dealing with some aspects of academic mathematics, it also showed how mathematics plays a crucial role in the world of work, in art and craft, in sculpture, in all societies, in games around the world, in mazes, in puzzles of many kinds, in the work of the magician, in computers, on postage stamps, in knots and in many other contexts. Most of these uses are not widely known, but the response of those who visited the exhibition showed that there was clearly a deep interest in what was available. Traditional school mathematics rarely deals at all with any of these contexts and mathematics teachers are not generally aware of the richness of the subject outside the material of the textbook.

One of the main attractions in the PopMaths Roadshow, particularly to the young, was, predictably, any activity that made use of microcomputers. In itself this comment raises important issues for mathematics education.

COMPUTERS

Both computers and calculators must be seen as extensions to a person's ability to do mathematics. There is always the possibility that a calculator will be used to perform trivial calculations which could have been done mentally, or that the computer will be used to perform a task in a laborious way, perhaps by listing all possibilities when it might have been more elegant to use a logical argument. The evidence, however, seems to suggest that regular users soon achieve a balance between the use of the machines and other methods. Where the balance is struck will depend on the individual. For those who find the simplest mathematical work difficult, the computer may be the key to providing access, for the first time, to achievement and success in the subject. For the ablest young people, the computer has the potential to allow them to develop their mathematical skills to levels far beyond those traditionally expected. The reality is that computer and calculator use is now a normal activity and it would seem apparent that one task of education is to ensure that people are able to make effective use of them at whatever level is appropriate for the individual.

Of particular interest to the mathematics teacher should be the existence now of an important sub-culture of young people who have high skills in computer use. It is tempting to be dismissive by claiming that children only play games with their home-owned computers, but some of the games have sophisticated mathematical aspects. Two contrasting examples will illuminate this point. One currently very popular game is BiaTetris (Klosters, 1990), in which the player has to manipulate shapes rapidly by translation and rotation in order to achieve a good score. The skills developed are similar to those that are developed in early geometrical work at school, but the context is highly motivating. A second example is of a flying simulation called 747 Flight Simulator (DACC Ltd, 1983), which requires the user to interpret dials, bearings, heights and climb rates in order to place an aircraft in space and bring it in to land in the dark. The skills of a good performer are considerable and include many that are mathematically significant at school level. Many young people are acquiring skills through their use of computers that need to be studied in much the same way as ethno-mathematics. The skills include those associated with the use of games and other packages as described above, and also quite advanced programming skills which use mathematical techniques. These computer skills that pupils bring with them to school are part of the ethno-mathematics of the developed society, and carry for the mathematics classroom a number of possibilities not yet really explored.

IMPLICATIONS FOR THE SCHOOLS

It would be wrong to expect too much of education, let alone mathematical education. The idea that mathematical education can have significant influences in developing a democratic society or in bringing about industrialization, as has been suggested by some writers, is preposterous. If society itself is not ready for particular types of developments, then they will probably not happen.

The subtlety of the interaction between humans and their environment makes it clear that a simplistic idea of cause and effect is inappropriate in any discussion of the role of education in general and of mathematical education in particular. There are, nevertheless, many ways in which mathematics is important in the total environment, and some of the links between the subject and society have been dealt with in this chapter. There would appear, however, to be no simple solutions for curriculum construction that take account of the reality of mathematics in all its forms. There are messages for the teacher and the curriculum developer in the way that mathematics is used in everyday activities, in the workplace and within every culture, but, as has been argued, the messages are difficult and it is notoriously hard to use what they convey to produce teaching materials. Perhaps the most important single message is that teachers need to be aware of the situation, to keep abreast of the way that mathematical usage in society develops and changes, and to bring to the curriculum personal attitudes and knowledge that are appropriate to the task of teaching a subject that is alive in a wide variety of forms.

REFERENCES

Archenhold, W. F., Wain, G. T. and Wood-Robinson, C. (1981) Integrating mathematics and science; cooperation, compromise or conflict. *Studies in Science Education* **8**, 119–26.

d'Ambrosio, U. (1985) Ethno mathematics and its place in the history and pedagogy of mathematics. *For the Learning of Mathematics* **5**, 44–8.

Bird, D. and Hiscox, M. (1981) *Mathematics in School and Employment: A Study of Liaison Activities*. London: Methuen Educational.

Bishop, A. J. (1988) *Mathematical Enculturation*. Dordrecht: Kluwer.

Cleeves, I. (1991) Who wants industry–education links anyway? *Mathematics in School* **20** (3), 36–7.

Cockcroft, W. H. (1982) *Mathematics Counts*. London: HMSO.

DACC Ltd (1983) *747 Flight Simulator*. Peterborough: Dr Soft.

Fitzgerald, A. F. (1981) *Mathematics in Employment (16–18)*. Bath: University of Bath.

Fitzgerald, A. F. (1985) *New Technology and Mathematics in Employment*. Birmingham: University of Birmingham Department of Curriculum Studies.

Gay, J. and Cole, M. (1967) *The New Mathematics in an Old Culture*. London: Holt, Rinehart and Winston.

Gerdes, P. (1988) On culture, geometrical thinking and mathematics education. *Educational Studies in Mathematics* **19** (2), 137–62.

Klosters, O. (1990) *BiaTetris*. Bergen: Brothers in Arm.

Mellin-Olsen, S. (1987) *The Politics of Mathematical Education*. Dordrecht: Reidel.

Orton, A. (ed.) (1985) *Studies in Mechanics Learning*. Leeds: University of Leeds Centre for Studies in Science and Mathematics Education.

Pinxten, R., van Dooren, I. and Harvey, F. (1983) *The Anthropology of Space*. Pennsylvania: University of Pennsylvania Press.

Pollak, H. O. (1979) The interaction between mathematics and other subjects. *New Trends in Mathematics Teaching, Volume IV*. Paris: UNESCO.

Rabinovitch, I. (1973) *Probability and Statistical Inference in Ancient and Medieval Jewish Literature*. Toronto: University of Toronto.

Sewell, B. (1981) *Use of Mathematics by Adults in Daily Life*. Leicester: Advisory Council for Adult and Continuing Education.

Simmons, J. (ed.) (1980) *The Education Dilemma*. Oxford: Pergamon Press.

Todhunter, I. (1865) *A History of the Mathematical Theory of Probability from the Time of Pascal to that of Laplace*. London: Macmillan.

Wain, G. T. (ed.) (1978) *Mathematical Education*. London: Van Nostrand Reinhold.

Wain, G. T. (1992) Reflections on the PopMaths Roadshow. Unpublished report. Available from The Centre for Studies in Science and Mathematics Education, University of Leeds.

Wilson, B. J. (1981) *Cultural Contexts of Science and Mathematics Education*. Leeds: University of Leeds Centre for Studies in Science Education.

Chapter 3

Learning Mathematics: Implications for Teaching

Anthony Orton

PROBLEMS OF LEARNING AND TEACHING

Teachers of mathematics quickly discover that learning does not take place automatically as a result even of what they consider their most outstanding lessons. This should not really surprise us, because adults also place their own interpretations on, and are selective in what they internalize from, conversations, experiences and events. Unfortunate misunderstandings in interactions between adults can arise all too easily, with consequences which range from the embarrassing to the fatal. School lessons judged as outstanding by an outsider might, in any case, often be performances which please both performer and spectator but which do not necessarily have much impact in terms of lasting learning. Yet quality of interaction between learner and environment, of which the teacher may form an important element, must, one suspects, have considerable effect on quality of learning, though direct evidence to prove this is very hard to find. Whatever definition of teaching is held, it is certainly an important and continuous responsibility of teachers to seek out and practise what they believe to be the most effective ways of promoting learning. All the evidence we have suggests that the issue is much more complicated than the notion conjured up by the current jargon of 'delivering' a curriculum to pupils, which is used so often nowadays in Britain. After all, when anything, whether concrete or abstract, is delivered at our doors it can be argued about or rejected, and even if accepted, it can subsequently lie untouched or even forgotten.

The search for best methods of teaching is one that has quietly continued since the dawn of civilization, and from time to time notable teachers have been recorded by history. Many years ago Highet (1951), for example, discussed what can be learned from some of the most effective teachers of the past. In terms of learning mathematics, the work of Froebel, Tillich and Montessori in earlier centuries continues to exert influence, and throughout the twentieth century there have been other successful innovators. Many searchers after improvement have looked to ideas emerging from educational psychology for guidance, or for confirmation that what they were already doing was best for their pupils. It could, indeed, be argued that it is successful and effective teaching practice that has led to the definition of some of what has subsequently become blessed with the description of 'theory'. Also, one particular problem has been ably summed up by Shulman (1970), in that 'mathematics educators have shown themselves especially adept at taking hold of conveniently available, psychological theories to buttress previously held instructional proclivities'. From time to time, it also has to be admitted, misguided attempts to improve learning have been introduced into schools, and it is only a small comfort to be able to record that it has often not been educators who have been responsible for these. Thus, a critical look at relevant theoretical ideas about learning must logically form a part of anyone's consideration of how the learning of mathematics is best fostered. Within one chapter of a book, however, it is necessary to be selective in what is considered and, in line with the intentions of the book, what space is available will be devoted largely to contemporary views and developments. For a more detailed consideration of alternative theories as they have been developed over the years and other aspects of learning mathematics, the reader should consult Orton (1992).

THE CONSTRUCTION OF MEANING

The simple arithmetical task:

$$23 \times 7$$

is one which even some 15-year-olds cannot complete without converting the question into an addition sum, a phenomenon which is usually a surprise to the teacher when it is first encountered. Naturally, it is very frustrating for us as teachers to see 23 written out seven times, often in column form rather than row form, because we feel entitled to believe that the multiplication algorithm should have been mastered many years earlier. In defence of their teachers of earlier years, it is well nigh certain that the algorithm has been taught, but the teaching has had little lasting effect on these particular pupils. In defence of such pupils, it is clear that they have learned something of mathematics; after all, they do appreciate that multiplication can be interpreted as repeated addition, they do have a degree of confidence about certain concepts relating to addition, and the addition procedure has been mastered. In other words, they have clearly mastered some mathematics but are just not yet at ease with the multiplication algorithm. At the other extreme, with more able pupils, it is not uncommon to find that the conversion of the task into:

$$230 - 69$$

is accepted without demur as an alternative procedure to the standard algorithm and can be carried out without recourse to pencil and paper. The point of this example is not to remind or demonstrate to readers that there exists a wide range of competency in mathematics within any age range, though that is in itself an important fact which teachers have to reckon with at all times; it is to draw attention to the frequent rejection of teacher-taught methods by some pupils. It seems that there is more to persuading pupils to use an algorithm than demonstrating how to carry out the steps of the procedure and giving plenty of opportunity to practise, a fact of life which many outside the teaching profession fail to realize. Unless pupils have an adequate understanding of a procedure, unless they have in some way made it their own, there is often some resistance to trying to use it. Pupils often benefit if they understand why a procedure works (relational understanding) as well as simply being able to carry out the procedure (instrumental understanding).

It should be noted here that, in discussing pupil competence, we can all glibly use words such as 'understanding', assuming that there is a uniformly held definition in the minds of those who hear or see our words. In fact, like all words, 'understanding' can mean different things to different people. Our use of any word does not convey our meaning, it merely calls up the meaning already held in the mind of the listener or reader, though this meaning is subject to change over years and it is possible that our use of the word might trigger amendment to the meaning previously held. In fact, we have all arrived at our own meaning of any word over a period of years, and although we feel able to relate it to and use it in the manifold contexts in which it is applicable, we have no idea what shades of alternative meaning are held by others. There are many parallels with learning mathematics here, which the ensuing discussion should reveal. In terms of the particular word in question, one helpful elaboration of the meaning of 'understanding' lies in the aforementioned distinction between instrumental understanding and relational understanding, widely used and discussed in mathematical education literature, for example in Skemp (1989). Contemporary literature might, in similar vein, also wish to distinguish between what is described as 'procedural knowledge' and as 'conceptual knowledge'.

So how do we assist pupils towards an adequate understanding of, for example, a particular algorithm like multiplication, if demonstrating it does not work? This question brings into sharp focus the issue of how learning takes place. There are no complete answers to the question, but there are justifiable and valid views, there are theories, and there is empirical evidence collected both by teachers over a lifetime of teaching and by educational researchers, often themselves teachers working within a research project or for a higher qualification. One important contemporary view is known as constructivism. In essence, constructivism is based on the view that, in the last resort, we all have to make sense of the world ourselves, and we continue to develop our understanding throughout life, just as we all come to our own definition of any particular word and continue to amend it. Such a straightforward claim may appear to some readers to be a truism, but the claim conceals contentious and difficult issues such as which version of constructivism is the truest (for there are alternatives), what the implications for teaching practice are of a belief in constructivism, and what the major difficulties are when teachers attempt to use methods which they believe are the best for enabling the construction of meaning. It is particularly important to scotch, from the outset, any equating of constructivism with 'free-for-all activity', 'discovery learning', 'child-centred education' and 'progressive education'. In any case, all of these terms, in addition to being used frequently in recent years in pejorative ways, themselves mean different things to different people and would require definition before use. The following statement about constructivism from Cobb *et al.* (1991) is quite categorical: 'constructivist teaching does not mean that "anything goes" or that the teacher gives up her authority and abrogates her wider societal obligations.' At a time when educational methods are under scrutiny from wider society and require careful justification, it is important to define precisely what is meant and what is not meant by constructivism.

The essence of constructivism, according to von Glasersfeld (1991), is that 'knowledge cannot simply be transferred ready-made from parent to child or from teacher to student but has to be actively built up by each learner in his or her own mind'. The development of constructivism owes much to the research and epistemology of Piaget (see for example Orton, 1992), but ideas have moved on. Constructivism does not inevitably imply that learners can only progress on their own. It does not mean there is nothing the teacher can do to assist in another's learning processes. Some would say it says nothing categorical about valid teaching methods and learning environments. It should not be equated with doing practical work, or using an investigatory approach, or involving pupils in discussion, or any one particular methodology. To many constructivists it does not rule out any form or style of teaching, rather it places enormous responsibility on the teacher. In short, it seems the teacher can still have defined goals, and can design tasks, assignments, problems, projects and other forms of learning that stimulate thought and mental activity, which are likely to help pupils towards appropriate achievements in mathematics. The teacher might indeed also frequently guide the activities of the pupils and their discussions with some degree of directness but only rarely through telling. In discussing constructivism in terms of implications for teaching, and whatever anyone's own particular beliefs about constructivism, the following summary by Richards (1991) is critical: 'Students will not become active learners by accident, but by *design*.' The concept of 'activity' and the implications of 'design' both require consideration.

VERSIONS OF CONSTRUCTIVISM

It is doubtful whether the statement from von Glasersfeld (above) is sufficient to define what might be called a theory of learning, but it does present something of a philosophical or epistemological stance. In further elaborations of what constructivism is, two hypotheses have emerged (see for example Lerman, 1989).

1. Knowledge is actively constructed by the learner, not passively received from the environment.
2. Coming to know is an adaptive process that organizes one's experiential world; it does not discover an independent, pre-existing world outside the mind of the knower.

The first hypothesis is relatively uncontroversial and is close to a restatement of the earlier quotation from von Glasersfeld, but it still needs further interpretation. It has been taken to imply that the emphasis in classrooms should be on mathematical activity not passive reception, as the quotation from Richards implies. On this matter von Glasersfeld comments, 'The notion that knowledge is the result of a learner's activity, rather than of the passive reception of information or instruction, goes back to Socrates.' But what is meant by mathematical activity? It might well be true that pupils frequently seem to appreciate a classroom buzzing with activity and in which they are engaged in practical tasks. It might well seem to us that learning through physical activity with concrete objects is often the most effective means of trying to promote the learning of particular ideas. This, however, might be an optimistic view, for the amount of learning taking place when apparatus is being used may not match our expectations. Also, there are other interpretations of 'activity'. It is important to realize that a critical objective is creative mental activity, and that the mind can be very active even when listening to a teacher or reading a book, or in other occupations which at first sight might appear passive. In other words, any interpretation of constructivism as activity-based learning is too simplistic because, for many people, this would certainly carry a notion of physical manipulation or movement. A constructivist view of learning may well not rule out any teaching technique in principle, and a variety of techniques may well be in regular use in a constructivist classroom, based on the needs of individual pupils. In fact, interaction might be considered an essential aspect of whatever is thought of as action. The optimum balance between concrete activity and mental activity within the mathematics curriculum is likely to vary according to age and capability of pupil, to nature and type of content, and to availability of suitable materials, anyway. What should be clear, however, is that the pure transmission mode of teaching, that is, the lecture method in any of its forms, is regarded by constructivists with suspicion and as generally ineffective and inadequate most of the time. Some pupils, some of the time, can perhaps cope for short periods with the direct transmission of ideas by a teacher, but the likelihood is that, within any class and at any particular moment, most of the pupils would not be benefiting. This style of teaching, with the incorporation of some questions and answers, has been widely used in mathematics lessons in the past, and is indeed often referred to as the 'traditional' method. For many of those who accept that constructivism underpins the best explanations we have of how real learning is maximized, transmission teaching is therefore basically unacceptable.

The second hypothesis, stated above, is much more radical. Indeed, those who accept both hypotheses have become known as 'radical constructivists'. According to Lochhead (1991), 'constructivism is [first] a statement about the nature of knowledge and its functional value to us.' The second hypothesis denies the existence of certain knowledge and raises questions about what any individual accepts as known. For the most radical constructivist there is no possibility of any certain knowledge about the world. The concept of 'understanding' thus once again emerges as problematical. In the words of Lerman (1989),

> if all understandings are private and individual constructions, no student behaviour will allow me to do anything other than make my own private construction about what the student 'understands' of my 'understanding' of the concept or idea in question.

There is a problem if we try to tie the notion of 'understanding' to the idea of certain and absolute mathematical concepts, for none of us can be sure what that absolute is, if it exists, never mind whether our understanding is the same as the absolute. There is not only a problem in knowing whether the understanding achieved by another is the same as ours, there is also the problem as to whether it is the same as anything absolute or objective. Indeed, in relation to the use of language as the only way we have of comparing our notions, we are also led to wonder whether it is ever possible to understand adequately what anyone else is saying. This radical view of knowledge is naturally in opposition to more objectivist views which do support the notion of the existence of agreed objective knowledge. It is also relevant here to introduce the concept of a 'consensual domain', which von Glasersfeld (1991) explains in this way:

> If . . . people look through distorting lenses and agree on what they see, this does not make what they see any more *real* – it merely means that on the basis of such agreements they can build up a consensus in certain areas of their subjective experiential worlds . . . one of the oldest [such areas] is the consensual domain of numbers.

The implications of the most radical of constructivist views for teaching are likely to be considerable.

Having tried to differentiate between weak and radical constructivism, we now turn to social constructivism or socio-constructivism. Here, an important feature is the role of social interaction and communication in assisting individuals to construct their own understanding. Cockcroft (1982) drew attention to the desirability of 'discussion between teacher and pupils and between pupils themselves' without saying a great deal about the purposes of such discussion in relation to how it helps learning, particularly between pupils. Cobb *et al.* (1991) said, 'social interactions between partners influence their mathematical activity and give rise to learning opportunities'. This seems to suggest pupils working in pairs, but the socio-constructivist might, of course, wish to use small groups of any appropriate size, according to the circumstances, in order to encourage social interaction and provide a forum in which questions are debated, ideas are discussed, and constructive criticism is the order of the day. The Department of Education and Science (DES, 1985) said, 'The quality of pupils' mathematical thinking as well as their ability to express themselves are considerably enhanced by discussion.' The larger the size of group, however, the more chance there is that some members of the group will play little or no part. None of this is new and unique to constructivism as a view of how learning might be enhanced through discussion, of course, and the issue of how pupils should be grouped is not one which lies wholly within constructivism (see Chapter 7). If there are any new insights from socio-constructivism in terms of the value of discussion, they are perhaps in the view that meaning, or understanding, is being to some extent actively negotiated through such discussion. In terms of radical views, if there is no absolute knowledge, only our own interpretation of the world, discussion at least allows the possibility of some mutual agreement within a group, the development of a consensual domain. Indeed, Solomon (1989) said it encourages the development of 'an essentially social being for whom knowing number involves entering into the social practices of its use'. And according to Balacheff (1991):

> Students have to learn mathematics as social knowledge; they are not free to choose the meanings they construct. These meanings must not only be efficient in solving problems, but they must also be coherent with those socially recognized. This condition is necessary for the future participation of students as adults in social activities.

Discussion with peers certainly can assist learning. The very articulation of ideas and views presents them for inspection, criticism and amendment both by the individual and by other members of the group. Thus discussion can help individuals to clarify their own notions, but it can also provide opportunities for the receipt of new ideas. In the value it places on social interaction, the socio-constructivist perspective thus perhaps owes more to the work of Vygotsky than to Piaget's view of the solitary knower who must construct meaning alone. The role of the teacher within a socio-constructivist approach is a sensitive one. According to Lochhead (1991):

> The primary goal . . . is to help students develop skills of constructing, evaluating and modifying concepts . . . The teacher's role therefore is to work to improve the quality of the discussions rather than to focus from the beginning on the 'correct' . . . answer.

The evidence we have suggests that many of us have great difficulty in not focusing on the correct answer when 'guiding' discussion; indeed, we often have great difficulty in holding back from direct interference and telling. Basically, we wish pupils to construct what we think they should construct. An important rider to discussion of socio-constructivism as the explanation of how learning best takes place is that it might seem to suggest that transfer or even transplantation of ideas from one pupil to another will happen. Transplants can be accepted and can make the host more whole, but they can also be rejected, and this inevitably often occurs in learning situations. Growth, however, takes place from within, and it seems to be important in any version of constructivism to acknowledge that a critical feature is what is invented or reinvented from within, with or without the benefit of social interaction. In fact, many who like to be thought of as constructivists would not consider what an avowed socio-constructivist believed to be radically different from their own beliefs.

An alternative contemporary view to constructivism as an explanation of how we learn, which also depends on interaction and argument within group situations, is sometimes described as the socio-cultural view. Socio-cultural theories appear to be concerned with transmitting meaning from one generation to the next as part of the process of acculturation. The view expressed by Solomon (1989) and quoted above might appear to be more socio-cultural than constructivist. The same conclusion might even be drawn from the statement by Balacheff (1991), though it was written in relation to constructivism. There is undoubtedly some common ground between socio-constructivism and socio-cultural views in that both accept the importance of collective activities, but the literature suggests there could also be important differences.

As a further postscript to this discussion of constructivism, it is relevant to draw attention to the idea of mental models. Johnson-Laird (1983) said:

> Human beings . . . do not apprehend the world directly; they possess only internal representation of it, because perception is the construction of a model of the world. They are unable to compare this perceptual representation directly with the world – it is their world.

The suggestion that we all form mental models of the environment and that new experiences are then subsequently interpreted in relation to the models we have already constructed appears to be relevant to discussion of different versions of constructivism. Newell and Simon (1972) were among the first to express views that, in relation to problem solving, the solver first constructs a representation of the 'problem space' and this then governs the way the encoding of information is carried out. So how we set about solving problems is less likely to involve logical thinking and is more likely to be based on our model of the situation, and this in itself is context dependent. What happens after that may well be a form of hypothesis testing. According to von Glasersfeld (1983):

> What determines the value of the conceptual structures is their experimental adequacy, their goodness of fit with experience, their viability as means for solving problems, among which is, of course, the never-ending problem of consistent organization that we call understanding.

MULTIPLICATION: A BRIEF CASE STUDY

Whether we should be concerned that some children cannot use a pencil and paper multiplication algorithm to find the answer to 23 × 7 is itself debatable. However, most teachers are both guided and constrained by a defined curriculum which, in the Western world and in countries whose curricula are based on those of the West, is almost certain to include the multiplication algorithm by pencil-and-paper methods, perhaps sub-divided into short and long versions. It is therefore important to address the issue of what the presence of this curriculum topic implies – what kinds of behaviour pupils should ultimately be able to demonstrate, to use the jargon of behaviourism. To some members of society, the objective is that pupils should be able to arrive at a correct answer to questions such as 23×7, by rote if necessary, and it should not take them long to do it. To others, the objective is more subtle and indirect and, it is often claimed, much more important. It is a greater understanding of number and arithmetic, a greater confidence in manipulating numbers which therefore inevitably includes competence with multiplication; it is observable progress in the development of mastery of the par-ticular domain of human intellectual activity known as arithmetic (or mathematics). Thus the hope might be that 23×7 can be comprehended and computed in many dif-ferent alternative forms, for example:

$$23 \times 7 = 20 \times 7 + 3 \ \times 7$$
$$= 23 \times 10 - 23 \ \times 3$$
$$= 12 \times 7 + 11 \times 7$$
$$= 42 + 42 + 77$$
$$= (25 - 2) \ (10 - 3)$$
$$= (20 + 3) \ (5 + 2)$$

There is such a multiplicity of alternatives like the above, even more with a question like 39×27, that mastery cannot be demonstrated simply through answering an exercise devised by the teacher or by a textbook author, for such exercises are always limited by space and time and constrained by the thinking of those who compiled them. The pupils themselves need to generate their own alternative forms of the question and must check them through. This is an activity which could be carried out partly on an individual basis and partly on a group basis. In any case, group activities inevitably generate within them many individual and temporarily private mental activities, as has been suggested earlier. The larger group is useful for widening the horizons of all members, for throwing up alternatives which not everyone has generated, and for promoting thought and discus-sion. The exploration of numbers and of ways in which numbers can be combined and operated on is a creative activity with a purpose. There may, of course, be problems with weaker pupils. Wood (1988) put it this way: 'Children who are mathematically gifted . . . often invent ingenious and *workable* methods for solving problems – methods that their teachers never envisaged. Children who *struggle* sometimes do the same thing: the problem is, their methods do not work.' If there are pupils who are incapable of exploring 23×7, then it could be argued that they are not incapable *per se*, but are simply not ready and may be able to explore simpler products, and that this would be more beneficial than attempting to use an algorithm which can only be imperfectly learned by rote. Readers may recognize some similarities with arguments for discovery learning, however we define it, propounded thirty years ago. The similarities and dif-ferences between constructivism and discovery learning do need to be understood.

It is sometimes said, and often by those whose minds are closed to teaching methods which attempt to encourage creative thinking among pupils when learning mathematics, that discovery learning is not justifiable, for there is no way that children can reinvent the whole of mathematics. In the hands of a skilful teacher, justifiable creativity can be encouraged without setting out to recreate mathematics, rather with the objective in mind of generating a deeper understanding of number and a comprehension of, for example, multiplication which is more likely to stand the test of time than when learned by rote. Such must have been the objectives of countless pioneers of former generations, both those whose work has been documented, like the proponents of structural apparatus for example, and many others who will forever remain anonymous. Such were surely the objectives of the proponents of discovery learning in the 1960s and 1970s, but like many other worthy attempts to rejuvenate the curriculum, discovery learning has been frequently misunderstood and often badly practised. In terms of differences between discovery learning and ways of learning suggested by constructivism, discovery learning was an attempt to persuade teachers to change their modes of operation within classrooms and might therefore have led to an exploration of multiplication like the one outlined above. Constructivism is a collection of beliefs about the way we learn which seem to lead one to discussions about the nature of knowledge. In one sense, therefore, there seems to be no relationship between the two. However, when a constructivist operates in the classroom it is possible that there will be many occasions when it looks like discovery learning all over again, because of the emphasis on the active recreation of knowledge. Hopefully, if it looks like discovery learning, it will be practised in the way which was always intended by such as Bruner (1960) and Biggs (1972). Neither constructivism nor discovery learning should lead to purposeless classroom activity which appears to have no structure and which goes nowhere. A closer look at classroom practice implied by the acceptance of constructivist views is now relevant.

CONSTRUCTIVIST TEACHING EXPERIMENTS

Although it should be clear by now that constructivism does not prescribe particular teaching methods, it is nevertheless relevant for us to try to discuss features of classroom practice and the problems which arise in attempting to implement constructivist beliefs in the classroom. It may, in fact, be fundamental to do this, given that teachers will not move away from 'traditional' methods, where clear guidance for so-called 'good' practice already exists, unless guidance as to how to implement constructivism also exists. In fact, Lochhead (1991), drawing attention to this need, has said, 'To date constructivist thinking has been more effective in describing what sorts of teaching will not work than in specifying what will.' The attitude of constructivists to transmission teaching and its rejection by many as being ineffective has already been mentioned, but what methods are to be encouraged, and do they work at least as well as, if not better than, traditional methods? These are critical questions given the pressure to return to basics with which teachers in some countries are having to contend. Simplistic interpretations of a return to basics are likely to include returning to traditional teaching methods, so alternative approaches to fostering effective learning will need to be seen to work. The acceptance of constructivist principles clearly suggests that pupils need to be provided with opportunities to construct their own mathematics for themselves, which the earlier example based on multiplication has attempted to illustrate. Large-scale experiments have, however, already taken place, and it is appropriate that attention should be drawn to some of these.

The work of Kamii (1985, 1989) records children reinventing arithmetic for themselves. The inspiration for this particular approach to promoting understanding through the pupils' own reconstructions of number rules is claimed to come from Piaget and the belief that children acquire mathematical knowledge not by internalizing rules imposed from outside but by construction from the inside through their own natural thinking abilities. When errors are committed, it is said, these arise because the children are thinking and not because they are careless. Thus the task of the teacher is not to try to correct from outside, but to create a situation in which the children will inevitably correct themselves. There is, however, also a strong hint in the evidence available that the basis of the methods used is, in the context of this chapter, best described as socio-constructivism. Children, it is claimed, construct mathematical knowledge more solidly when they are encouraged to defend their ideas within a group or even a whole class. The knowledge which is acquired is then that which the group agrees should be accepted.

The beginnings of arithmetic for the children involved lies in learning about numbers and number combinations through playing dice games, which leads to the memorization of number bonds without any direct teaching, worksheets or other kinds of reinforcement from the teacher. It could be said that this is, at least in part, learning by rote, but here the method works well, the use of games providing essential motivation. Subsequently, numerical questions posed by the teacher form the focus for class discussion of answers suggested by individual children, sometimes privately to the teacher, sometimes openly to the whole class. Thus, in one sense, the teacher is very firmly in control, and contrary to many of the ways in which pupils work in groups nowadays, the questions are presented on the chalkboard to the whole class. Drill and worksheets do not play any part at this stage either, and pencil and paper are not available, so the emphasis is on 'mental' rather than 'mechanical' arithmetic. Number tasks are not as a matter of general practice invented by the children but are part of a carefully devised sequence of the teacher's making which is intended to provide for and enable progression in capability and knowledge. The critical part which the teacher plays is thus apparent. When answers are offered by individual pupils, others declare whether they agree or disagree. Disagreement among children leads to other offers which, if the class consensus is one of agreement, leads to the pupils being encouraged to explain how they arrived at the answer which has received the approval of the group. This is a very revealing aspect of the method, for there is normally variety in how the children have constructed their solutions to a particular problem, so children have the opportunity to reflect on alternative approaches. For example, the task:

$$\begin{array}{r} 2\ 7 \\ -\ \ 1\ 8 \\ \hline \end{array}$$

might lead to some children thinking of it in terms of $28 - 18 - 1$, others as $27 - 17 - 1$, others as $20 - 10 + 7 - 8$, others as $20 - 10 - (8 - 7)$, and so on. In terms of generalities, allowing children to find their own way through mathematics reveals that most children deal with the tens before the units, this perhaps going some way to explaining why attempting to teach an algorithm which runs counter to what at this stage seems to be an apparently natural tendency runs into difficulty. It also reveals that some number combinations are remembered more easily than others, for example doubles like $6 + 6$ are remembered more easily than $5 + 6$, which explains why, when dealing with

$$\begin{array}{r} 2\ 6 \\ +\quad\ 7 \\ \hline \end{array}$$

many children will deal with the units as $(6 + 6) + 1$. At a much later stage, evidence has been collected which shows quite young children eventually not only coping with methods of multiplying two 2-digit numbers together, but also just about arriving at the formal multiplication algorithm on their own.

No doubt some readers will be inclined to voice a criticism here, namely that this is all very well but it takes too much time and there is a syllabus to get through. It does take considerable time when children are allowed to work things out for themselves, but one legitimate rejoinder is whether time is saved in the long run, for so much time is currently spent in mathematics lessons on reteaching and providing routine practice on ideas which do not seem to have been mastered first, second, or even third time round when taught in the 'traditional' manner. If the extra time leads to genuine progression in mastery of mathematics, if ideas are understood more firmly and lastingly by allowing children greater opportunity to work things out for themselves, this is a more worthwhile gain than any teacher's satisfaction in completing the syllabus. At least, this is the argument, and it is one which has been with us for a very long time. Constructivism is breathing new life into attempts to teach for understanding, and it must be hoped that it will not founder because of half-baked or timid interpretations which produce worse results than traditional teaching methods do. More evidence from work like that of Kamii would, however, help us all to convince ourselves and thus have the courage to change.

The work of Bell *et al.* (1989, for example) in conflict teaching in the 12–14 age range may be taken as a second example of a successful experiment in constructivist approaches to teaching mathematics. In this style of teaching, there has first to be a phase when particular beliefs held by pupils have to be revealed as inadequate or inaccurate. According to Bell, subsequent direct instruction, that is, 'intensive teaching, focussed sharply on known misconceptions', proved comparatively unsuccessful in the respect that 'there was very little transfer to points not so strongly focussed upon, even though they were dependent on the same general concepts'. Conflict teaching proved much more successful in this respect. The process thus begins with the exposure of difficulties and misunderstandings, perhaps through tried and tested diagnostic questions. Once disagreement has occurred, small-group discussion is set up to argue out the differences and come to a socially agreed view which can then be taken to the whole class for further discussion. The small-group work is important because it is more likely to expose the extent of any misconceptions, and because pupils are both more prepared to talk and are put under more pressure to talk the smaller the group. Although Bell does not claim that the method is based on constructivist principles, and in fact claims no support from any particular theory or epistemology, it seems clear that conflict teaching forms one approach which constructivists might be pleased to claim as a technique worth advocating. Cognitive conflict has, in fact, been claimed by Underhill (1991) to be one of the two major mechanisms which motivate learning and which induce reflective activity, which itself stimulates the cognitive restructuring required within the process of constructing knowledge and understanding.

It is, of course, typical of education that there are legitimate criticisms which can be levelled at the kind of group teaching methods described above. In their various ways, such methods put pressure on pupils to conform to a particular view, so it is possible that meaning has only been negotiated to the extent that some pupils will finally admit that something must be correct solely because a majority seem to believe it. For the sake of social survival, pupils might on the surface accept the group view, but at the same time their confidence might have been undermined if they believe they are the only member who still does not really understand. They may consequently become further convinced that they are no good at mathematics, and so their attitude might deteriorate. Thus the outcome from a conflict situation could, in theory, be very unfortunate. As an example, no matter how able and how well taught, most students will at some time expand:

$$(a + b)^2$$

as

$$a^2 + b^2$$

and not as

$$a^2 + b^2 + 2ab.$$

If they do not understand why the last alternative is what the teacher seems to believe and what the rest of the group appear to accept, they may well decide to conform without really understanding. In the last resort, however, they will only understand when that understanding comes from within, when they can appreciate how the identity fits in and makes sense within the greater scheme of things known as mathematics, when the result has a firm place in the connected network of ideas in the mind. Social pressure might well serve to pull individuals up short and force them to rethink, but nothing can inject or transplant into the mind from without. Thus, whatever the value of group methods and of cognitive conflict methods for many pupils, learning remains a very individual activity. Is socio-constructivism anything more than individual construction within a social setting? Indeed, whatever value is to be obtained from pupils working in groups, the ultimate authority for the pupil does not lie within the group, it resides with the teacher, and beyond that with the literature and the community of mathematicians. The construction of meaning might well be the most effective way to learn, but the thorough student might still feel the need to check not only with other members of the group but with a higher authority.

Even prior to the development and clear exposition of constructivism as a concept, mathematics teaching, in many countries, had begun to move away from a reliance only on the 'traditional' transmission mode, despite the fact that many people outside teaching, and some inside, have viewed such moves with suspicion (and herein lie the seeds of possible future backlash against attempts to put constructivism into practice). Investigational methods, problem solving and extended projects are now clearly built into the curriculum in many countries at various stages of pupils' education. There are reasons for these developments which perhaps have nothing to do with constructivism, but it is possible that constructivism might provide a clearer rationale for such teaching methods, and also provide teachers with clearer guidelines as to procedures to adopt when practising the methods. If the major intention becomes that of providing pupils with opportunities to construct their own understanding of mathematics, and if this is to be carried out at least partly on a social rather than entirely individual basis, investigational and project work provide excellent opportunities for this but need to be structured in such ways as to attain these ends better. The balance between individual effort, small-group effort, teacher intervention and whole-class discussion can be guided by the underlying philosophy that the pupils need to come to a better understanding of mathematics, through their own creative efforts, but in the environment in which conclusions need social approval. Again, the onus is on the teacher, and great skill is required, far more so than by transmission teaching. Further discussion of constructivist teaching will be found in von Glasersfeld (1991) and in *Educational Studies in Mathematics* **23**(5).

CONSTRUCTIVISM IN OUR CLASSROOMS

If constructivism does not provide us with a learning theory and does not of itself dictate to us what our teaching methods should be, it is relevant to look in rather more detail at which of the methods in widespread use in mathematics classrooms in the past, and indeed today, are legitimate to a contemporary constructivist. Here, we are again in difficulty, for there are various interpretations of constructivism and there is, as yet, little agreement as to how classroom practice is guided by any particular version.

At one extreme, it seems one can say that no method is ruled out, but at the other one might find the view that construction must imply mental and frequently concrete activity, and thus apparatus is often essential and the general climate of the classroom needs to be one of investigation, inquiry and discovery. The traditional and ubiquitous whole-class teaching method which has become known as transmission teaching has been fairly categorically rejected by many constructivists as being ineffective for most of the time for most of the class, as we have already seen, but individual children often need only the transmission of a quick answer to a question to promote their cognitive development, and a further question from the teacher is not only unneccesary, it is a frustration. As a result of being told something, another mental cog can be fitted into place in the wheels of the mind. It therefore does not seem legitimate to assume that constructivism suggests you should never tell any child anything. Behaviourism has influenced teaching methods around the world throughout this century. It led to emphasis on drill and practice in mathematics in the belief that 'practice makes perfect', and the classroom experience of many teachers will perhaps lead them to believe that there should still be a place for practice, that children sometimes need mental practice just as they need motor practice when learning to ride a bicycle or to swim. It is interesting, however, that practice is said to be excluded in Kamii's experiments. Rote learning, that is learning by heart and without meaning, has often been associated, perhaps unfairly, with behaviourism, and has certainly been a traditional method of school teaching. However, rote learning is accepted as a necessary feature of the approach to learning mathematics associated with Kamii, but it is not forced, it takes place in a natural and enjoyable way. Many people have successfully and happily learned aspects of mathematics without fully understanding at the time. On the other hand, many others have come to reject mathematics because they did not understand enough – there was too little in the way of connections between networks in the mind.

So it really is very difficult to rule out any method; but attention is inevitably drawn to the uniqueness of the abilities, capabilities and needs of each individual learner. Ideally, the teaching method should be the right one for each child at any particular moment; it must be relevant to their existing level of knowledge and understanding (Ausubel, 1968). In summary, however, constructivism would seem to argue against the widespread class use of traditional methods because these are less successful in helping individual children than more individually based methods. Constructivism is certainly being interpreted as advocating discovery and inquiry-based learning, incorporating opportunities for discussion and the negotiation and exchange of ideas.

A problem which then immediately emerges is that teaching methods which may appear to be suggested when one tries to put constructivism into practice have already been tried and, some would say, have been found wanting. One example is discovery learning and another rather more specific one is the use of structural apparatus. The purpose of apparatus such as that devised by Stern, Cuisenaire and Dienes is to place into the hands of children wooden or plastic objects which implicitly contain a structure that is the concrete analogy of the abstractions within the mathematics to be learned. Thus Stern and Cuisenaire provide slightly different variations intended to introduce addition, subtraction, multiplication and division (with possible extensions into subsequent mathematical ideas), the Dienes Multibase Arithmetic Blocks provide a basis for learning about place-value, a domain of mathematics with which even many older pupils struggle, and the Dienes Logical Blocks can underpin some comparatively advanced mathematical concepts concerned with sets, logic, relations and functions. Other practical and concrete approaches to mathematical concepts which perhaps do not so closely reflect or are not so closely analogous to the mathematics as structural apparatus are also often advocated as good ways to help children over their difficulties.

The theory of Piaget has sometimes been held responsible for our enthusiasm for concrete approaches, and concrete approaches certainly also fit within the general ethos of discovery learning, though the advantages of using equipment in mathematics were documented by Froebel as long ago as the early 1800s. The problem is that apparatus has, in general, not proved to be as successful in helping many pupils as was hoped. For one thing, pupils do not always see or make the connections between the apparatus and the mathematics, an issue which has been discussed by Hart (1989). Constructivism, however, does not of itself advocate the use of concrete approaches, it simply suggests that children must construct their own meaning. It is our assumption, based to some extent on experience, that concrete apparatus will automatically help, so we cannot look to constructivism but need to look elsewhere for reasons for any assumed failure when we hoped for so much from apparatus. Nor should we assume, because we are sometimes disappointed with the effectiveness of apparatus of any kind, that this justifies transmission teaching.

In learning mechanics, Williams (1985) found that the use of apparatus helped a great deal. It should perhaps be stated that there is a difference between structural apparatus which sets out to provide concrete support for the extraction of ideas which are modelled by the structure and other apparatus which contains no such exact match with ideas. The apparatus used by Williams was in no sense structural but it did provide practical experience. It mostly helped the students because it challenged preconceptions, it revealed beliefs which were not borne out when the apparatus was used (and which were also, of course, contrary to those of objective mathematical knowledge as understood by the teacher), and it thus promoted discussion, further controlled experiment and the exchange of opinions which ultimately led to change of belief. The whole process from the revelation of alternative conceptions to the coming to an agreed view had to run its course, and this took time. Often, the use of apparatus in schools is curtailed by the teacher when it is judged either that enough time has been spent or that the majority of pupils have learned the knowledge or skill in question. Of course, it is always difficult for a teacher to know when there is no longer any need to provide apparatus. Sometimes the use of particular apparatus is associated with worksheets and becomes almost behaviouristic in its implementation. Often the use of apparatus does not demand that pupils think, discuss, argue and conclude. The success in the mechanics context was not solely due to the apparatus and was certainly not associated with filling in worksheets; it was due to allowing students the time as well as the opportunity to construct their own understanding. It was perhaps also due to the resolution of cognitive conflict within a social context. There can be no guarantee that younger pupils would eventually ever be able to use their experiences with Multibase Arithmetic Blocks, say, to come to a better understanding of place-value in the same way that the mechanics students eventually came to the conclusions hoped for by the teacher. That surely is compatible with, rather than a failure of, constructivism.

As regards using equipment, one can only suggest that too much is sometimes expected, that it is perhaps the case that the equipment is merely a vehicle which can provide both analogous concrete experience and opportunity for group discussion and the social construction of meaning. In many cases of reported failure, the focus has perhaps been too much on the provision of appropriate equipment and not enough on what goes on in the minds of the pupils. Indeed, some would say that the concrete embodiments of abstract ideas perhaps only directly convey the ideas to those who already understand them. It is important not to lose sight of the Piagetian view that it is the activity which is to be mathematized, not the equipment itself, nor to lose sight of the importance of discussion and argument, both between teacher and pupils and between pupils themselves. More research is needed into approaches to learning mathematics which might be described as socio-constructivist, both with and without concrete apparatus.

As regards discovery learning, similar conclusions are legitimate. Setting up what is intended to be a discovery learning situation is naturally not automatically going to guarantee learning, and may be an abysmal failure. Situations which introduce cognitive conflict and the opportunities for the negotiation of meaning within a group are closer to what might be needed when one sets out to implement constructivism, and these, too, need to be researched extensively. This is not easy to do when one believes that the constraints of extensive curriculum content are severe and when there is no encouragement to look for better ways of teaching.

The idea of concept mapping (see Orton, 1992) is one that can assist pupils in both checking and consolidating in a constructivist way what it is that they have learned. At the nodes of a concept map (network) might be items of knowledge, or concepts, or anything which a child remembers from lessons. Linking the nodes, lines can be drawn to show connections or relationships of any kind which the child believes exists. Figure 3.1 shows a pupil's representation of what had been learned about triangles, together

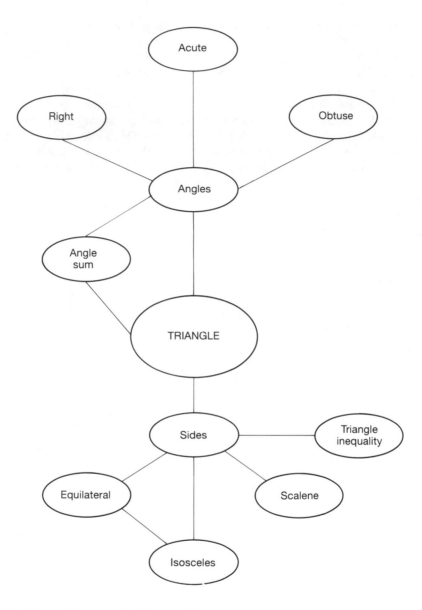

Figure 3.1   *A concept map produced by a pupil*

with the perceived connections. The constructivist point of view would be that it is better for children to construct their own representation of what they think they have learned than for the teacher to tell them what they ought to have learned. The diagram is then available for others to challenge and thus improve. An example of a map constructed by two pupils working cooperatively within the Wakefield Open Learning Project (see Chapter 9) is shown at Figure 3.2 (original spellings retained).

Another version of concept mapping has recently been suggested by Entrekin (1992),

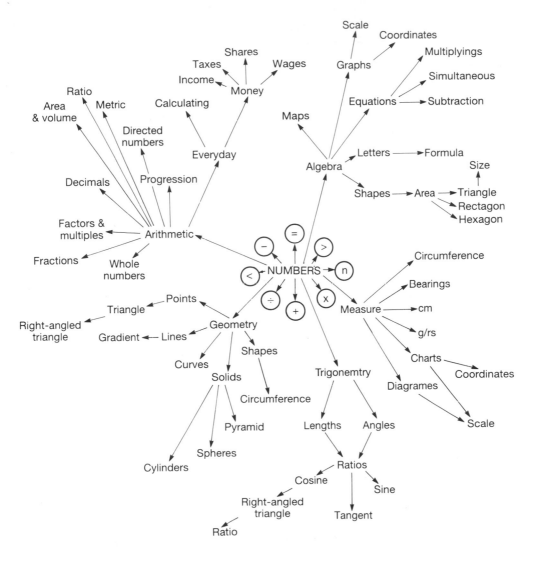

Figure 3.2 *A concept map produced by two pupils*

and this she calls 'mathematical mind mapping'. The basic idea is to begin a teaching sequence by asking for a word or phrase which best described what, say, yesterday's lesson had been about. This word or phrase might then be recorded on the chalkboard. Subsequent questions could be based on, 'And what did we discover about it?', or possibly, 'And what else?', with the aim of building up a map of what is in pupils' minds. This can clearly lead to class discussion, debate and argument, with the ultimate stage being either an agreed best map, or each pupil's own individual best map, of a subject area, showing how separate nodes are related to a main node and to each other. Maps completed individually after some early discussion can then lead to consideration of who has the best, and why, perhaps in a small group situation. An example adapted from Entrekin is shown in Figure 3.3. The choice of boxes is up to the pupils.

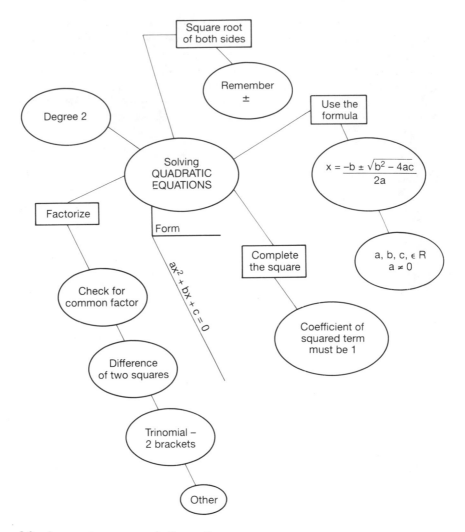

Figure 3.3   *A concept map on quadratic equations*

Some mathematical topics might lend themselves to students writing their own sections for a 'textbook'. An example from advanced statistics concerns the theorem that distributions based on sums and differences of two given distributions have mean equal to the sum (or difference) of the two means and variance equal to the sum of the two variances. This can be 'proved' by students, to their great delight. Initial investigation with simple numbers will suggest what the results are, and the students can then be set to work to produce algebraic proofs. The draft proofs produced in this way can then be debated and discussed and ultimately written out in a format which is as good as is found in any textbook. In this way a theorem has been constructed and not delivered, with the likelihood that it will not only be better remembered, but will always be used correctly. Here, however, we are clearly talking of able pupils. At the other extreme we have to acknowledge that there are weak pupils who may have very different needs. Not enough is known about the extent to which such pupils can construct their own understanding. The first reaction of many teachers is that such pupils need very clear instructions and a great deal of help, but the quality of learning is still often not good. Some hold the view that constructivist approaches need to be tried out with weak pupils as well, because the success rate can hardly be worse than it is with traditional transmission and practice. Underhill (1991) has described several more teaching techniques which can assist teachers to move towards styles of teaching which better promote construction.

SOME DIFFICULTIES

Many of the problems associated with putting constructivism into practice, if there is such a thing, certainly predate constructivism. Nevertheless, it is relevant to consider some of the issues, otherwise what is perceived to be advocated by constructivists may perhaps be judged unrealistic. One obvious problem which many of us will have already experienced is that anything other than transmission teaching is inclined to make us feel we are not really teaching. This is exacerbated by the view of teaching held by the general public, and is not helped if children claim at home that their teacher never tells them anything! It is a particularly acute problem for new teachers, who often have the view, perhaps the result of their own experiences as a pupil, that the fundamental basis of teaching is explaining to pupils, and sometimes come into teaching with the firm conviction that they will become good teachers because they are good at explaining things to others. Pupils, also, are inclined to think highly of teachers who 'can explain things clearly', and so it is reasonable to assume that this perception of the role of the teacher is carried through to when they themselves become teachers. Constructivism undermines the view that powers of explanation are important, unless it is the pupil explaining to the teacher, or perhaps to other pupils in a group situation. The ability to explain carefully, clearly and succinctly is certainly a very useful attribute for a teacher to have, but if pupils are to be allowed to construct their own knowledge, it is an attribute which will often have to be held in check. So is it ever allowable to use this skill, and if so when? Is it ever allowable to tell?

New teachers perhaps have to contend with a further problem in that teacher training usually emphasizes careful lesson preparation, and whatever the views held by teacher trainers about how children learn (and if anything teacher trainers have a reputation for being progressive rather than traditional), the structure of lessons advocated by trainers usually seems closer to behaviourist principles, from objectives to measured outcomes, than constructivist. In fact, new teachers desperately need as much security as can be provided, and traditional teaching methods are usually perceived as offering the greatest security there is in a very insecure occupation. Class control, for example, particularly for a new teacher, is generally said to be more difficult when methods other than transmission are being attempted. Of course, to some extent the comparative lack of security of methods other than transmission is also worrying to experienced teachers. It is clearly necessary to develop methods of providing confidence and support for teachers if there is to be a greater move away from transmission teaching.

In terms of teacher intervention, there are other problems too. Given free rein, pupils will frequently explore in directions which the teacher knows are either unhelpful or outside the content defined by the curriculum. Should pupils be stopped, in these circumstances? When should they be stopped? Can we risk them not working out for themselves that they are not going anywhere? Can we risk their displeasure with us if they find that we knew they were not going anywhere and deliberately chose not to tell them? What if the pupils have strayed into a domain and appear to be motivated and making good progress but the teacher knows nothing of this particular domain? Is it acceptable to admit that pupils know more than their teacher in any area of mathematics? New teachers in particular, but not solely, find this very hard to do. And even if the pupils are working within content bounds which the teacher knows are appropriate in terms of the syllabus, is there a time when the teacher should step in and curtail exploration, perhaps in order to hold a class discussion, perhaps with the intention of clarifying what has been learned? Is it allowable for the teacher to aim to close an episode, and if so what are the acceptable or best ways of doing this? Is summing up allowed? Decisions about intervention must also depend on how much progress any individual pupil is making. How do we recognize when learning is or is not taking place, for without this knowledge we cannot be sure if and when we should intervene?

When working to a prescribed syllabus, there is often a tendency to try to keep all the pupils in the class together. Traditional teaching methods tend to give the impression that the entire class of pupils forms a homogeneous unit in terms of cognitive development, and that all the pupils are ready to move on to a new idea or topic at the same time. Allowing children to generate their own knowledge and construct their own understanding on an individual basis quickly reveals the falsity of the view that there is such a thing as a homogeneous group. So what should one do about the slower pupils, and indeed about the quicker pupils? There are a number of alternatives, ranging from allowing all pupils to progress at their own pace, no matter how slowly, to curtailing by reverting to transmission at some chosen moment. One possible middle road is to have pupils working in groups so that slower pupils do at least become persuaded that there is group acceptance of an idea and they too had better accept it if they are to survive. Another possible problem is that individual pupils themselves might have preferred modes of learning which are at variance with what the teacher prefers. This is a problem which can arise when the teacher's preferred mode is transmission, of course, but it is also very common for pupils from time to time to express a preference for being told something by the teacher. How should the teacher react to such a request? Some pupils, furthermore, believe that working within a group slows them down. What seems certain is that individual teachers wishing to come to terms with constructivist views will have to work out for themselves what constructivism means for them and how to practise it in their own classroom, otherwise the problems described above cannot be faced. If constructivism is a valid description of how learning has to take place, then we certainly all need to construct our own meaning for it. This is clearly best achieved by reflection on the issues and possibilities, and it takes time.

The concern that the prescribed curriculum will not have been completed by the end of the year is a very real one for many teachers. The concept of completing any particular chunk of content is, indeed, hardly compatible with constructivist views of learning, but that does nothing to remove very natural anxieties. The most that can be achieved, if it is felt that the syllabus is an unavoidable constraint, is clearly some sort of compromise. Relevant issues here have already been alluded to in discussing what to do when pupils go off in unexpected or unintended directions. Furthermore, the wider view of curriculum includes the means of assessment, and some prescribed curricula now incorporate prescribed forms of testing. Many teachers would say that approaches to learning based on constructivist beliefs do not sit easily alongside the kinds of tests and examinations which pupils have to write. Any particular examination paper is devised to match what whoever set the paper thinks should have been constructed. It can only imperfectly reveal what has actually been constructed. In some parts of the world, modes of examining are changing, and individual project and investigatory work is being examined despite some resistance, which comes partly from within education but mainly from without. So far, however, the kind of cooperation between pupils encouraged by socio-constructivism is considered to be incompatible with assessment procedures.

Other aspects which many consider to form a part of the wider view of what a curriculum is might be of just as much concern. For example, what comprises the best learning environment when one believes that pupils construct their own mathematical knowledge? The environment needs to be rich in opportunities for learning, but what does this mean? How does one set about providing the best possible environment? And how does one take account of the fact that many of us work better to deadlines and within defined constraints? As we have seen, one powerful constructivist mode of instruction which is to do with the kind of learning environment provided is group discussion, often in the form of some type of cooperative problem solving. However, Cobb *et al.* (1991) have said: 'It should be clear that, for the constructivist, substantive mathematical learning is a problem-solving process . . . [but] the general notion that problems can be given ready-made to students is highly questionable.' Problems defined by the pupils themselves might sometimes arise but in most classrooms teacher-devised problems will be more normal. As regards issues concerned with discussion in groups, there are many, and these are considered in Chapter 7.

There are also problems which might be described as being associated with cultural differences or with differences of schooling practices around the world. One of the problems of introducing teaching methods which appear to diminish the direct teaching role and authority of the teacher is that we all, to a greater or lesser extent, live within a cultural milieu which has traditionally encouraged the view that the teacher is the authority and ultimate source of knowledge and wisdom, at least within school. Thus, for some countries around the world it is an enormous problem to move away from 'traditional' teaching methods. Children are expected to respect and listen to their elders without question, they expect to be told what to do and to be instructed in the ways of the social group, they do not expect to be placed in the position of having to be creative, and they do not readily accommodate to a role in which they may debate with and even question the view of the teacher. In such societies, teaching methods which attempt to encourage children to develop their own understanding and construct their own knowledge may take time to develop, but must not be considered impossible. All around the world, social relationships and behaviour are subject to change as we all 'move with the times'. In Britain, and many other Western countries, the problem of changing teaching methods may not be a very difficult one. Here, society has changed to such an extent that the questioning of authority on the part of the young is commonplace, so any problems in adjusting to new ways of learning mathematics may be just that – problems of adjustment on the part of both teacher and pupil. In a constructivist environment, the problem is that the teacher has to develop a new kind of authority – one which is founded on earning respect rather than having it accorded automatically by society, so again we see that the role of the constructivist teacher becomes even more demanding. Another problem for which the solution may be much more elusive is that not enough is known about how the ways in which mathematics is best learned vary according to the culture, and this suggests that constructivism in practice will need to be interpreted according to cultural diversity.

There is no doubt that the adoption of methods of teaching which reflect constructivist beliefs present difficulties, but effective teaching has always been, and probably always will be, hard work. It is appropriate to leave the last word to Wood (1988):

> The perspective I have adopted on the nature of knowledge and its relation to formal systems of thinking . . . precludes an approach to teaching that is based on universal and invariant 'steps' and 'stages' . . . Rather, it invites interaction, negotiation and the *shared construction* of experiences . . . The only way to avoid the formation of entrenched misconceptions is through discussion and interaction. A trouble shared, in mathematical discourse, may become a problem solved.

REFERENCES

Ausubel, D. P. (1968) *Educational Psychology: A Cognitive View*. New York: Holt, Rinehart and Winston.

Balacheff, N. (1991) Treatment of refutations: aspects of the complexity of a constructivist approach to mathematics learning. In E. von Glasersfeld (ed.), *Radical Constructivism in Mathematics Education*. Dordrecht: Kluwer.

Bell, A., Swan, M., Onslow, B., Pratt, K. and Purdy, D. (1989) *Diagnostic Teaching for Long Term Learning*. Nottingham: Shell Centre for Mathematical Education.

Biggs, E. E. (1972) Investigational methods. In L. R. Chapman (ed.), *The Process of Learning Mathematics*. Oxford: Pergamon Press.

Bruner, J. S. (1960) *The Process of Education*. Cambridge, MA: Harvard University Press.

Cobb, P., Wood, T. and Yackel, E. (1991) A constructivist approach to second grade mathematics. In E. von Glasersfeld (ed.), *Radical Constructivism in Mathematics Education*. Dordrecht: Kluwer.

Cockcroft, W. H. (1982) *Mathematics Counts*. London: HMSO.

DES (1985) *Mathematics from 5 to 16*. London: HMSO.

Entrekin, V. S. (1992) Mathematical mind mapping. *The Mathematics Teacher* **85**(6), 444–5.

Hart, K. (1989) There is little connection. In P. Ernest (ed.), *Mathematics Teaching: The State of the Art*. Lewes: Falmer Press.

Highet, G. (1951) *The Art of Teaching*. London: Methuen.

Johnson-Laird, P. N. (1983) *Mental Models*. Cambridge: Cambridge University Press.

Kamii, C. K. with DeClark, G. (1985) *Young Children Reinvent Arithmetic: Implications of Piaget's Theory*. New York: Teachers College Press.

Kamii, C. K. with Joseph, L. L. (1989) *Young Children Continue to Reinvent Arithmetic – 2nd Grade: Implications of Piaget's Theory*. New York: Teachers College Press.

Lerman, S. (1989) Constructivism, mathematics and mathematics education. *Educational Studies in Mathematics* **20**(2), 211–23.

Lochhead, J. (1991) Making math mean. In E. von Glasersfeld (ed.), *Radical Constructivism in Mathematics Education*. Dordrecht: Kluwer.

Newell, A. and Simon, H. A. (1972) *Human Problem Solving*. Englewood Cliffs, NJ: Prentice-Hall.

Orton, A. (1992) *Learning Mathematics: Issues, Theory and Classroom Practice*, 2nd Edition. London: Cassell.

Richards, J. (1991) Mathematical discussions. In E. von Glasersfeld (ed.), *Radical Constructivism in Mathematics Education*. Dordrecht: Kluwer.

Shulman, L. S. (1970) Psychology and mathematics education. In E. G. Begle (ed.), *Mathematics Education*. Chicago: NSSE.

Skemp, R. R. (1989) *Mathematics in the Primary School*. London: Routledge.

Solomon, Y. (1989) *The Practice of Mathematics*. London: Routledge.

Underhill, R. G. (1991) Two layers of constructivist curricular interaction. In E. von Glasersfeld (ed.), *Radical Constructivism in Mathematics Education*. Dordrecht: Kluwer.

von Glasersfeld, E. (1983) Learning as a constructive activity. In J. Bergeron and N. Herscovics (eds), *Proceedings of Fifth Annual Meeting*. PME-NA.

von Glasersfeld, E. (ed.) (1991) *Radical Constructivism in Mathematics Education*. Dordrecht: Kluwer.

Williams, J. S. (1985) Using equipment in teaching mechanics. In A. Orton (ed.), *Studies in Mechanics Learning*. University of Leeds: Centre for Studies in Science and Mathematics Education.

Wood, D. (1988) *How Children Think and Learn*. Oxford: Blackwell.

Chapter 4

The National Curriculum in Mathematics

Tom Roper and Dave Carter

INTRODUCTION

Mathematics is a core subject in the school curriculum of all countries. The National Curriculum of England and Wales (Mathematics) (DES, 1991), instituted by the Education Reform Act, 1988 (ERA), represents the legally binding entitlement to mathematics education of all pupils aged 5 to 16 within the state education sector. Similar, but not identical, arrangements appertain in Northern Ireland and Scotland. This curriculum has five Attainment Targets (ATs):

- AT1 (Ma1): using and applying mathematics;
- AT2 (Ma2): number;
- AT3 (Ma3): algebra;
- AT4 (Ma4): shape and space;
- AT5 (Ma5): handling data.

For each of these attainment targets, there are ten levels through which the pupil may progress according to previous attainment and ability. These levels are derived from the model of progression proposed in the Task Group on Assessment and Testing (TGAT) Report (DES, 1987) and accepted by the Secretary of State for Education. This model applies to all National Curriculum subjects and remarkably, for such an overarching model, has no research evidence quoted in the Report to support it.

For each level in each Attainment Target, statements which comprise the Programme of Study (PoS) are provided. These represent what is to be taught, under the heading 'Pupils should engage in activities which involve:'. Also provided are Statements of Attainment (SoA) . These state what pupils should be able to do as a result of taking part in the PoS, under the heading, 'Pupils should be able to:'. For each SoA, examples are given to exemplify and amplify the SoA, under the heading, 'Pupils could:'. Only the Programmes of Study and the Statements of Attainment are enshrined in law.

Assessment of pupils' attainment within the National Curriculum takes place at ages 7, 11, 14 and 16, termed Key Stages (KS) 1 to 4 respectively. The levels of attainment and PoS designated at each KS are respectively 1–3, 2–6, 3–8 and 4–10. The assessments, Standard Attainment Tests (SATs), are devised by agencies under contract to the School Examinations and Assessment Council (SEAC), the appointed body responsible for the assessment of the National Curriculum. Assessment of levels outside those designated at each Key Stage does occur; for example, the KS3 Mathematics and Science assessments, piloted in 1992, covered all the levels from 1 to 10. The responsibility for the curriculum itself lies with the National Curriculum Council (NCC). It is likely that NCC and SEAC will be combined to form one body in the near future (see the *Times Educational Supplement*, 22 January 1993). While ATs 2–5 are defined upon what can be seen as separate content areas of mathematics, they are meant to be linked by Ma1, which is process based (see Chapter 1). To use the analogy of weaving, 'Using and applying mathematics' is the weft to the warp formed by the other four ATs.

Having described the basic legal structure of Mathematics in the National Curriculum of England and Wales and the way in which the elements of it are intended to interact with each other, we now consider its recent historical antecedents and the developments or changes in the thinking about mathematics education which the introduction of it has brought about.

THE LINEAGE OF MATHEMATICS IN THE NATIONAL CURRICULUM

Mathematics, while clearly being part of wider curriculum development, has had a development of its own. It is entirely possible that this development has been one of the major influences in the development of the whole of the National Curriculum. Following the so-called Great Debate about Education, initiated by Prime Minister Callaghan through his famous speech at Ruskin College, Oxford, in 1976, a Committee of Inquiry was set up into the teaching of mathematics in schools in 1978. The report of this Committee of Inquiry was published by the Department of Education and Science (DES) in 1982 under the title *Mathematics Counts*, and is usually referred to as the Cockcroft Report, after its chairman, Professor Wilfred Cockcroft.

This report was welcomed by the majority of mathematics educators in the United Kingdom. It embodied within its recommendations many of the latest developments being practised in the classroom. Paragraph 243 summarizes the teaching practices approved by the Cockcroft Committee as follows:

Mathematics teaching at all levels should include opportunities for
- exposition by the teacher;
- discussion between teacher and pupils and between pupils themselves;
- appropriate practical work;
- consolidation and practice of fundamental skills and routines;
- problem solving, including the applications of mathematics to everyday situations;
- investigational work.

However, the report placed these recommendations as to what should happen in the classroom within a structure and philosophy of mathematics education throughout the years of compulsory schooling. The basis of this structure was that pupils should be successful at mathematics, they should not be exposed to ideas that they could not understand nor to so many ideas that they did not have the time to come to grips with them, and they should not be taking examinations in which they could not obtain a reasonable proportion of the marks available. The idea of passing an examination with a mark of 40 per cent or less was an anathema. From this premise it was argued that syllabuses needed to be constructed from the bottom up, not primarily based upon what higher-ability pupils were capable of achieving. Thus the construction of examination syllabuses at 16+ should not be primarily based upon the starting point for A-level (Advanced level), taken at 18+, nor should A-level be based upon the starting point for degree-level studies. Rather, the syllabuses at 16+ should be based upon a consideration of what pupils might need as a basic minimum for everyday life, for the world of work and as a basis for future study, informed by a knowledge of what children could do and understand, provided by such large-scale research studies as the Assessment of Performance Unit (APU) (1980a, 1980b, 1981) and the Concepts in Secondary Mathematics and Science (CSMS) (Hart, 1981). An immediate consequence of such a structure is differentiated syllabuses. The Cockcroft Report provided the starting point for such syllabuses by drawing up a Foundation List, which it considered to be a minimum entitlement for all pupils.

Sir Wilfred Cockcroft was appointed as the first Chairman of the new Secondary Examinations Council (SEC) with a brief to reform the secondary examination system at 16+. SEC put forward proposals for a new General Certificate of Secondary Education (GCSE) designed to replace the dual system of Ordinary level General Certificate of Education (GCE) and the Certificate of Secondary Education (CSE). It is worthy of note here that there was at this time a body charged with curriculum development already in existence, namely the Schools' Curriculum and Development Council. However, in the reforms of the curriculum which stemmed from the implementation of GCSE, it was the examining body, SEC, which led the way. That there is competition between the two bodies currently responsible for curriculum and assessment, NCC and SEAC, and that the proposed amalgamation referred to earlier is seen as SEAC taking over NCC, is not surprising considering earlier events.

The examination boards responsible for GCE and CSE came together to form regionally based examining groups. These new groups put forward syllabuses based upon the subject-specific criteria issued by SEC and the Cockcroft Report (Cockcroft, 1982). As a result they show an expected degree of uniformity. All the groups put forward schemes which included the assessment of work completed throughout the two-year course from ages 14 to 16, which was required to be largely of an investigative and practical nature, as suggested by paragraph 243 of the Cockcroft Report. Each group offered three levels of syllabus and stepped examination papers. Their syllabuses for the lowest level were based upon the Cockcroft Foundation List. It is only at the highest level that any real differences in syllabus content can be found, and these can hardly be seen as significant. Each level of syllabus is a proper subset of the syllabus at the next higher level, and so the syllabuses can be represented as shown in Figure 4.1. Therefore within mathematics a *de facto* national curriculum existed for pupils aged 14 to 16 with very similar assessment arrangements across all the examining groups, even prior to the introduction of the National Curriculum.

GCSE in mathematics can be seen as representing the establishment of several developments in thinking about mathematics education which differ from or are significant extensions of the thinking of the 'modern mathematics' era of the 1960s and early 1970s. Mathematics in the National Curriculum, as the successor to GCSE in the secondary sector and the means of introducing a formal mathematics curriculum

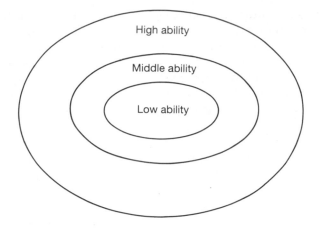

Figure 4.1 *Levels of examination syllabuses*

into the primary sector, marks the further extension of, or change in, these areas of development.

PROGRESSION AND PSYCHOLOGY

In the 'modern mathematics' era the majority of pupils whose mathematics education was within that tradition encountered the same ideas; for example, middle-and high-ability students were formally introduced to matrices and topology in the form of networks. This was reflected in the watered-down GCE syllabuses which many pupils met as CSE syllabuses at the upper secondary level. At the lower secondary and primary levels, the philosophical and psychological standpoint was that the basic structures of mathematics were similar to the basic structures of the thinking process and hence all pupils should have access to them. Hence Piaget (1973) states that

> Thus, having established the continuity between the spontaneous actions of the child and his reflexive thought, it can be seen from this that the essential notions which characterize modern mathematics are much closer to the structures of 'natural' thought than are the concepts used in traditional mathematics.

However, Thom (1973), seeking to argue against this position, restates it as follows:

> The psycho-pedagogues, aware of the vagueness of their doctrinal position, believed that they had found the key to their problems in the assertions of logicians and formalist mathematicians. Since it was acknowledged that the progression of mathematical thought was modelled by those great formal schemata that are the structures – structures of sets and logic, algebraic structures, topological structures – teaching the child, at an early enough age, the definition and the use of these structures would suffice to give him easy access to contemporary mathematical theories.

On the other hand, GCSE offers different mathematics for pupils of different abilities. There are some mathematical ideas that experience suggests some students are not capable of understanding and using. Therefore these students should not meet these ideas, but should gain a better and fuller understanding of ideas which they are capable of grasping. For example, calculus was not to be formally introduced before the age of 16, whereas matrices and vectors could be treated in a formal way with pupils of higher ability below the age of 16, but only in a qualitative way with pupils of lower ability. However, within each of the differentiated syllabuses, the ordering of the ideas was up to the teacher. Furthermore, below the age of 14, there was considerable freedom as to what ideas might be introduced and when, given the constraints of preparing for a public examination at 16+.

The National Curriculum takes this idea a stage further. The necessity of laying out the mathematics curriculum along the lines of the TGAT model of ten levels, coupled with the formal assessment required at regular intervals, implies that there is a strict hierarchy of difficulty and that, until the lower levels have been mastered, the upper levels should not be attempted. Since the curriculum is a legal requirement, enforced via Act of Parliament across the whole of the age range of compulsory schooling, the professional responsibility of teachers in the matter of what is taught and when is considerably reduced. Thankfully, the way that any particular topic is taught is still the responsibility of the teacher. However, even this may be under challenge. Such a straitjacket of progression or hierarchy of difficulty comes at a time when there has been a gradual movement towards seeing mathematics more as an interconnected web of topics and subtopics which do not have a strict universal hierarchy, though there may well be local hierarchies within individual topic areas. The writings of such authors as Davis and Hersh (1983, 1988) have done much to promote this view of mathematics.

However, a stronger influence than this, but interacting with it, has been the growing importance of two paradigms. The first is the constructivist view of learning (see Chapter 3). With the emphasis which this paradigm places upon the construction of knowledge by the individual and the resulting personal and, almost certainly, idiosyncratic understanding of that knowledge, the idea of a global hierarchy of difficulty has little to commend it. The second is what might be termed 'the problem-solving' paradigm. Within this paradigm the essence of mathematics is solving problems via mathematical thinking. The process element of mathematics is given as much emphasis as the content element if not more (see Chapter 1). Thus the paradigm can become a method of teaching and learning mathematics, motivating the study of the content in order to solve real problems. Since what is a real problem depends upon individuals, their own understanding of it, attitude towards it and interest in it, this paradigm is often seen as being the classroom approach best suited to promoting the constructivist paradigm. Such relativistic and individually orientated paradigms do not fit easily with an absolutist and monolithic curriculum for all.

Progression through the curriculum will undoubtedly be gauged by results attained in the national assessments, although teacher assessment will advise on level of entry. The problem to be highlighted here is that progression depends upon the interpretation placed upon the various Statements of Attainment for the purposes of assessment. Thus, for example, within the National Curriculum, an operation and its inverse are often combined in an SoA at the same level. This does not take into account that the inverse operation is generally regarded by experienced teachers as being harder than the operation itself.

An example is to be found in AT4, SoA 6(d):

Demonstrate that they know and can use the formulae for finding the areas and circumferences of circles.

The examples given to illustrate what is required of the student in order to demonstrate that this SoA has been attained are two in number, one of which requires the basic operation, the other the inverse. In any test of this SoA what exactly will be tested, the operation or its inverse? If both are tested, will the student have to get both correct or only one? Will the teacher know beforehand in order to ensure the correct level of entry? Will the same criterion be maintained from one year to the next or will it vary? In the case of this last question, it is a distinct possibility that the criterion will vary over time within a Key Stage and across Key Stages. This is because of the intention of the government to award contracts for the construction of SATs at the end of different Key Stages to different bodies and for limited periods of time. Will all the bodies involved, present and future, share identical interpretations of the SoA? This is clearly a very important question when considering the notion of progression through the National Curriculum. However, there is a further problem which is perhaps even more difficult to resolve. If we suppose that all the bodies have identical interpretations of the SoA, then what will they expect from different students of different ages in response to test questions? The original intentions were that the mathematics should be embedded in contexts suited to the experience and maturity of each age group. We must point to research which clearly shows that context has a great influence upon the response rate (APU, 1986) and ask how this is to be taken into account.

To take it into account by recourse to decontextualized questions is not an answer; the problem remains. For example, AT3 SoA 6(b) states:

Solve simple equations.

The corresponding Programme of Study statement is:

Solving linear equations; solving simple polynomial equations by 'trial and improvement' methods.

and it is exemplified by:

Solve: $3x + 4 = 10 - x$.
Solve equations such as $x^2 = 5$ and $x^3 = 20$ by 'trial and improvement' using calculator.

This SoA will be tested at KS2 and KS3, but what will the test questions look like? At KS2, it is unlikely the pupils will have been taught the necessary techniques in the manipulation of algebraic expressions. If this is acknowledged then questions such as

$$6x + 1 = 7 \quad \text{or} \quad 5x - 3 = 7$$

might be set. However because these questions can be answered by 'guess a number' methods, are they really comparable with the example given in the curriculum, which requires a degree of manipulation to determine the answer? If they are to be regarded as comparable then we are in effect saying that a child aged 11, at KS2, has attained the SoA by solving $5x - 3 = 7$, but that a child aged 14, at KS3, has still only attained the same SoA by solving, say, $3x + 15 = 10 - 5x$, a question which requires the manipulation of symbols and the use of negative and rational numbers. Such a conclusion appears to be unfair from the point of view of the KS3 child, and yet how are KS2 children to show that they can attain this statement if they have not been taught the necessary content and techniques?

Such contradictions occur at all the Key Stages and their resolution is important because they affect progression and readiness. If they remain unresolved, it is difficult to see how teachers working with pupils in the succeeding Key Stage can place any reliance upon the results from the previous stage in planning their courses. In the example given above, the teacher of a year 7 or year 8 child who had been awarded AT3, 6(b), would need to take great care to establish precisely what that meant in terms of the child's readiness to embark upon subsequent and related programmes of study. Should the state of readiness be at the stage of solving $5x - 3 = 7$ only, then much of the work implicit within that Programme of Study statement and SoA remains to be taught. Progress through the levels may therefore not be as rapid as parents, governors and senior staff of a school might suppose on the information provided by the previous Key Stage assessment. One of the main findings of Her Majesty's Inspectors (HMI, 1992) is that 'Little progress was made during the year on assessment and recording. Information gained from assessment, particularly in Key Stage 2 and Key Stage 3, was not used sufficiently to plan further work for pupils.' It is easy to appreciate why this should occur when the interpretation of the information is vague and fraught with difficulties. If future progression is governed by present knowledge, many teachers will automatically err upon the side of making absolutely sure that their pupils have at least been exposed to that which they consider essential for the next stage.

What we have tried to show here is that progression through the National Curriculum is dependent upon the interpretation of the SoA by a large number of individuals and statutory bodies; these SoA are relative and not absolute.

The Graded Assessment in Mathematics project (GAIM, 1988) and work by Denvir *et al.* (1987) have shown that for a given SoA, suitably exemplified, two experienced teachers should be able to assess independently whether or not a student has achieved that SoA. Noss *et al.* (1989) argue very strongly against this conclusion on many of the grounds used above. However, the crucial difference as to why the findings may not be transferable to the national situation is that teachers achieved high degrees of consistency of interpretation of statements only because of a great deal of direct support from the originators of the SoA. To be fair, Denvir *et al.* (1987) do caution against extending their work without sufficient support for the teachers involved.

METHODS OF DELIVERY

Let us first dispense with the word delivery. Furniture and groceries are delivered; a curriculum is taught, it is not deposited on the doorstep. Conscious efforts, involving professional skills and training, are made to ensure that as much of what is taught as possible is learned. The current jargon of education seems deliberately intended to reduce teaching to a trade rather than maintain it as that which it is, a profession.

Secondly, the founders of the School Mathematics Project (SMP) recognized the importance of having a syllabus for a formal examination as a means of ensuring that new content material would be taught (Thwaites, 1972). Sir Wilfred Cockcroft, as chairman of SEC, was able to enforce the recommendations of his report, particularly paragraph 243, this time by control of the whole examination system and not just a particular syllabus. Furthermore, an attempt was made to try to change not only what was taught, but how it was taught through the introduction of the assessment of process skills, the basis of investigational and practical work.

Again, the National Curriculum takes this line of thinking a stage further in imposing both syllabus and assessment across the whole of the age range of compulsory schooling. Politically, much has been made of the need to return to basics. Just what these basics are has never been defined, neither has the point at which they were abandoned. Presumably the curriculum as published and enshrined in the 1988 Act, and refined by the Education (National Curriculum) (Attainment Targets and Programmes of Study in Mathematics) Order, 1991, represents these basics and the assessment is there to ensure that they are taught. This particular point is worth stressing, since the opinion is sometimes advanced that the National Curriculum would be all right if the assessment requirements were reduced or dropped altogether. This is simply not possible. The aim of the exercise is to ensure that this particular variant of the basics is actually taught. This cannot be guaranteed without assessment at regular intervals. Hence the National Curriculum cannot exist without the associated assessment.

However, while the legislation determines what is to be taught and how it is to be examined, it steps back from the Cockcroft position of attempting to tell teachers how they should teach. Instead it offers advice through the Non-statutory Guidance associated with the 1989 version of the National Curriculum and still valid by direct reference in the Non-statutory Guidance for the 1991 version of the curriculum. The advice offered amounts to two variations on the problem-solving paradigm based upon the Attainment Target Ma1, 'Using and applying mathematics'. However, despite some materials published by NCC (see, for example, 1991a, 1991b, 1992a, 1992b) to assist in this respect, it is clear from HMI (1991, 1992) that in fact these approaches are being neglected. Teachers are concentrating upon the other four Attainment Targets, which are all content-focused rather than process-orientated.

Given the assessment arrangements, this is hardly surprising. First and foremost, ATs 2–5 will be externally assessed; only AT1 will be formally assessed by the teacher. This carries the clear message that what counts is the content, not the process. Secondly, there is disquiet and uncertainty as to what the Statements of Attainment in AT1 mean and how to use them to assess pupils' work. This has been seen in the popularity of a local in-service course, Assessment of Using and Applying Mathematics at KS3, at the University of Leeds. Some 300 teachers from around 200 schools have attended this course at the time of writing. The dissatisfaction with NCC and SEAC publications concerning Ma1, and their ineffectiveness, is also manifest. Our experience in running these courses suggests that teachers at the secondary level are content to use one-off investigations or practical pieces of work to assess Ma1 and use teaching time to drill content skills, despite encouragement to move towards integrating Ma1 into their teaching. HMI (1991, 1992) suggest that this is also the case in primary schools where the introduction of Data handling and probability (Ma5), as well as the need to cover more content in the traditional fields of Number (Ma2) and Algebra (Ma3), has led to increasing pressure to move in this direction.

The declared intention of government to publish school results without reference to the pupils' starting points – that is, the raw scores as opposed to the 'added value' – will increase the pressure to teach only content, ensuring that pupils are well versed in that which is to be formally assessed. This could easily signify a return to methods of 'delivery' not dissimilar to those used in the era of payment by results introduced as a result of the Newcastle Commission of 1858 (see for example Hamilton, 1976).

A question to be raised here is whether those parts of the National Curriculum which are not assessed will be given due emphasis in the classroom. The pilot KS3 SATs of June 1992 tested roughly two-thirds of the SoA, the other third being acknowledged as being unassessable (Cornforth, 1992) as they involved work with computers and other technology not available in the examination room. Figures from an SMP computer survey (Harrison, 1992) suggest that, by and large, mathematics departments are using the software identified by the National Curriculum. However, an average of 46.7 pupils to a computer in mathematics classes suggests that coverage may well be superficial, an inference supported by the fact that two-thirds of the departments who responded replied 'No' to the question, 'Do you feel that your department is sufficiently equipped in terms of hardware, software and expertise in their use, to be able to deliver the computer-related ATs?' Assessment can assure those who need assurance that the prescribed curriculum is being taught, but only those parts of it which are assessed. To assume otherwise flies in the face of common sense. Equally, while there is neither prescription nor proscription concerning how the curriculum is to be taught, there are pressures from the assessment schedule which teachers may find extremely difficult to resist.

A final irony is that the ideal pathway through the curriculum is one which is prepared for the individual student. Individualized learning schemes and supported self-study should be blossoming (see Chapter 9). However, personal experience would suggest that the opposite is in fact happening. Teachers will only purchase texts and schemes which they feel will assist them in teaching the prescribed curriculum. The apparent popularity of texts which closely resemble Ordinary level GCE texts of some thirty years ago suggests a return to a form of basics which many teachers and pupils had hoped had gone for ever.

IN-SERVICE TRAINING

Howson *et al.* (1981) examine several modern mathematics projects and curriculum development initiatives of the 1950s and 1960s under various headings. One of the main conclusions that they draw is the necessity for good dissemination of a curriculum if its ideals are to be fulfilled in the classroom. Teachers must not only be aware of the new material that is to be taught but must have an understanding of and sympathy with the means by which it is to be taught and the philosophy that underpins the project. The same authors also point out that the further away from the centre of the developments teachers are, the more likely it is that the messages of the project will be corrupted.

The SMP and Midlands Mathematical Experiment (MME), in developing their new Ordinary level courses during the late 1960s and early 1970s, rapidly discovered that an annual conference was insufficient to cope with the increasing number of schools wanting to take up the new ideas and materials. They therefore began to organize a series of regional courses which ran for several years at regular venues. The main aim appeared to be to inform teachers of the new material contained within the syllabuses and introduce them to the way that this material was presented within the textbooks of the projects. The most recent SMP 11–16 material has taken this in-service (INSET) element even further by establishing local user groups which meet regularly to work together as well as to be informed about the latest developments. Thus within a geographical region there will be a shared understanding of the material and the aims of the project. By being part of an organized group, teachers can exert a much greater influence on the project than they could as individuals.

The Cockcroft Report (Cockcroft, 1982) also recognized the need for teachers to be made aware of the developments that were being put forward within the Report and being held up as good practice. Among the recommendations that it advanced was the setting up within local education authorities (LEAs) of teams of advisory teachers. These teachers, selected for their abilities in the classroom, were to be seconded from their schools for a period of one, two or three years in order to go into the mathematics classrooms of other schools and work with the teachers there to improve the teaching of mathematics. They would also provide INSET for their LEAs in mathematics, but away from the classroom, where there was time for teachers to reflect upon what it was that was being discussed. Once the period of secondment had elapsed, the advisory teachers were to return to the teaching staff of the LEA and be replaced by others. The recommendation was accepted, the government provided funds for the secondment of these teachers, and LEAs set up teams of advisory teachers to help in the dissemination of good practice across the full age range of compulsory schooling.

Kinder *et al.* (1991) report, among other things, upon the various modes of operation of the advisory teachers and would seem to indicate that the advisory teachers were successful in doing what they were established to do. Many of those chosen to be advisory teachers were either appointed to head of department posts when their period of secondment was over or obtained posts as LEA advisors or inspectors in mathematics. While the report notes that some advisory teachers did not welcome the return to the classroom, feeling that they had acquired skills and abilities that would not be exercised to the full in the classroom, they certainly constitute a valuable resource within the teaching staff of any LEA.

A second recommendation of the Cockcroft Report was the establishment of coordinators of mathematics in primary schools. The report of Stow and Foxman (1988), while offering some criticism of the workload and number of different roles within schools that these coordinators were called upon to take up, notes once again the success of an initiative suggested by the Cockcroft Report.

Crucial to both of these initiatives is the idea of working directly with teachers and pupils in the classroom. The advisory teachers, as visitors to the classroom, were able to bring with them advice, expertise and knowledge to be used to encourage teachers to change their existing practice; the mathematics coordinators, as resident experts, were in a position to offer daily support and encouragement to colleagues. The work of the Low Attainers in Mathematics Project (Ahmed, 1987) confirms that lasting changes can be made in classroom practice in this way and that much greater commitment and cooperation is achieved when teachers are engaged at a personal level in the developments which will affect them and their pupils.

The Education Reform Act 1988 (ERA) established local management of schools (LMS). LMS has caused the LEAs to devolve most of their funds direct to the schools themselves. Hence many LEAs no longer have the finances to support large teams of advisory teachers and so INSET provision by the LEAs has been reduced. This has occurred in tandem with the introduction of the National Curriculum. It is strange that at a time of major curriculum change, when a network of experienced in-service personnel working with teachers in their own classrooms would have been of enormous benefit in introducing the National Curriculum, this network has been crucially weakened, and in some areas of the country would seem to have disappeared altogether.

On the other hand, money has been found by central government, channelled via the LEAS, to support the enhancement of the mathematical knowledge of primary-school teachers through the medium of 20-day courses. The government provides 60 per cent of the funding, the LEA the remainder. Within these courses, teaching at a higher education (HE) institution is integrated with practice in the classroom. Such courses have necessarily concentrated upon increasing the mathematical competence of primary-school teachers, rather than teaching methods, since the teachers are being placed in a position of being expected, via the National Curriculum, to teach more topics (and different ones) in mathematics than they taught previously. These courses, however, are not usually available to all primary-school teachers. Those who are selected by the LEAs are usually the mathematics coordinators, and they are then expected to share their increased knowledge with the rest of their school's staff, and in some instances with other schools in their locality.

The devolvement of funds directly to schools includes money for INSET provision. Schools are to decide upon their own needs and provide for them. They can therefore choose to go and buy whatever is available from sources such as their LEA and local HE institutions. However, the amount of these monies appears to be small in relation to the schools' perceived needs. Hence many school mathematics departments are providing their own INSET. They are aided in this by a wealth of material produced by the NCC and the SEAC, for example NCC (1991a, 1991b, 1992a, 1992b) and SEAC (1992a, 1992b). The former provides documentation containing activities for INSET work, either within the individual department or for use by other providers, which are designed to help the teacher come to terms with the curriculum and have suggestions as to how to implement the curriculum in the classroom. The latter provides equivalent documentation to assist teachers in the assessment of the curriculum.

These documents are sent to every state-maintained school in the country. They have frequently been written under contract by staff of HE institutions and are often based upon a project that has worked very closely with a number of schools. However, while that close relationship between the writers and the schools involved has helped all immediately concerned to come to grips with the subject of a publication and develop a mutual understanding, simply sending the publication out to other schools does not guarantee that its message will be received as intended. Indeed, it does not guarantee that it will be read in any detail at all, let alone studied or used for INSET.

The effectiveness of this form of INSET has yet to be evaluated. Nevertheless, there must be a degree of pessimism about the outcome of such an evaluation in the light of the various reports quoted above. This is confirmed by HMI (1992): 'The response of teachers to the original assessment INSET packs from the School Examinations and Assessment Council (SEAC) was negative; some schools ignored it, others found it confusing.' The lack of personal involvement of the majority of teachers in the construction of the materials and the issuing of edicts from a central source would seem to imply that the schools which are on the periphery of the development, in this case the majority, will interpret the National Curriculum only in terms of its most highly visible component, assessment. Should this happen, then it would seem inevitable that much of what has been seen as good practice in the mathematics classroom, stemming from the Cockcroft Report, will be discarded in favour of more instrumental methods such as drill and practice.

RESOURCE IMPLICATIONS

The introduction of any new curriculum places demands upon resources, both those that are already in existence and those which are required specifically to support the latest developments. The National Curriculum has prompted the introduction of many new texts, all of which purport to 'cover' the curriculum. It has also seen the alignment of existing texts, workcard schemes and individualized learning schemes with the PoS and the SoA and the making good of any deficiencies in order to sustain their place in the market. Where such schemes are produced by a school, a great deal of work has had to go into 'modernizing' the materials.

However, the National Curriculum makes new demands upon the resources of the schools; for example, the use of the computer and certain associated software is written into the curriculum. As noted above, Harrison (1992) suggests that teachers *are* teaching the computer-related segments of the National Curriculum. However, it is clear that there is considerable need for there to be more hardware available in the mathematics classroom or directly available for mathematics teaching. Such hardware is often there within the school but, as Harrison reports, it is in specially equipped computer suites. This puts demands upon the abilities of teachers to plan ahead sufficiently to enable such facilities to be booked well in advance. While this is easily said, it is often very difficult to do, and not just because of the difficulties inherent in planning and organizing lessons to fit to a strict timetable over a long period of time. The children being taught will almost certainly have a central role in determining the pace and order of future lessons, based upon their reactions to the previous one.

If there is easy and open access to a resource, then it will be well used; if access is restricted, then either alternatives will be found or the problem will be avoided altogether. But it is not just human nature which intrudes. Information technology (IT) is also a central plank of the National Curriculum as a whole. This in itself places additional demands upon the facilities available in the school, making them even more inaccessible. Mathematics cannot claim any computers that are within the school as its own, but certainly the minimum provision would seem to be one per mathematics classroom. However, this requires the dedication of specific classrooms to mathematics, a trend which the Cockcroft Report sought to encourage, or the networking of all classrooms as remote stations with access to a central server. Many schools have moved in this direction already.

There are, however, two alternative ideas to this which might prove to be adequate stop-gaps, and there may also be intrinsic merit in developing these further. In the first instance there exists the possibility of accessing hardware that is not the school's, but in fact belongs to the pupils or their families. The Mathematical Association (1989) developed a series of homework tasks to be done on the pupil's own home micro-computer, the tasks being offered as an alternative to more standard homework tasks. This idea might well be extendable into the areas of the curriculum where the computer is essential. Pupils who do not have access to a microcomputer at home could be offered guaranteed access at school. While such a scheme might be attacked as being potentially socially divisive, it offers the possibility of greater use of microcomputers by all the children and encourages the effective use of home-based machines. Problems may also exist in terms of availability of software in the home; it may not be possessed by the home, and if it is, there is the question of the compatibility of it with that of the school. Forms of licensing agreements could be used in some way to overcome this problem.

In the second instance, it may be possible to have the IT staff teaching the mechanics of the use of the various types of software, if they do not already do so, and incorporating into their planning examples and problems provided by the mathematics department. Such a solution has dangers in that by passing over to other teachers the responsibility for the computer-based work, mathematics staff who are computer 'shy' will have no cause to become better acquainted with the mathematical tools of our time. Those who already use them clearly feel the need for more knowledge (see Harrison, 1992). Every effort must be made to ensure that all mathematics staff are fully competent in this area. After all, there are times within mathematics lessons when it is useful if the teacher can turn to a computer. Given the pace of current developments in this area, 'useful' may soon become 'essential' (see Chapter 2).

The extension of certain specific branches of mathematics into the primary sector, for example statistics and probability, makes heavy demands upon teachers. While schemes such as the 20-day courses noted above have been put into operation with the intention of meeting these demands, there will still be pressure upon that most valuable of resources, suitably qualified staff. This pressure will increase if some of the current thinking from the NCC (1993) and the views expressed by Alexander *et al.* (1992) are put into action. These views have been interpreted as being in favour of the teaching of mathematics in the later years of the primary school by specialist rather than generalist teachers. It is not at all clear whether there are sufficient of these teachers within primary education. One effect could be to restrict the influence of the mathematics coordinator to classes of older pupils within the school and curtail the role of support for other staff. HMI (1991) indicate quite clearly that this has already occurred in some schools: 'Some coordinators made little impact, and often if they were teaching Key Stage 2 classes they had only minimal effect on colleagues teaching in Year 1.' Timetable complications are almost certain to arise concerning the organization of the rest of the teachers and children while such specialist lessons take place.

There will be pressure on time, which primary teachers will need in order to become accustomed to the new topics which they will be required to teach. Mathematics coordinators will need time to train their colleagues. But when will this be done? Other subjects new to the primary curriculum in their detail and scope, if not in actual fact, will demand an allocation from the time available.

Within the secondary sector, the compulsion to include subjects within the curriculum for all pupils at all times within certain age ranges has seen the reduction in time allocated to mathematics in some schools, coupled in some cases with a reorganization of that time across year groups in order to maximize its effective use. Thus within a notional allocation of hours per week throughout the five years of secondary schooling, year 7 may well be given a greater share than year 8, working on the argument that it is better to win them for mathematics while they are young and keen.

There are some signs that the pressure on time may ease, especially if the number of subjects which are compulsory is reduced. However, there are several cross-curricular themes, for example health education, whose supporters would like to see them taught in quite specific time slots. Any easing of the statutory subject-specific demands will therefore not automatically benefit those subjects which have given up time in the recent past. Other areas of the curriculum may be able to show greater or more urgent cause to benefit from the available time.

CONCLUSION

The National Curriculum of England and Wales was introduced at great speed; as a result the mathematics curriculum has already been revised once. Mathematics also often leads the way in terms of assessment. The first KS3 papers were in mathematics and science, and mathematics will be in the vanguard at KS2. As a result, mathematics teaching has seen many changes in recent years, some immediate and contradictory to the gradual shifts in practice of years prior to the changes. Others have been more in the nature of continuations of previous developments, but have taken those developments to extremes.

There is now a desperate need for a quiet time when teachers at all levels can come to grips with what is expected of them. Indeed, at the joint conference of the Mathematical Association (MA) and the Association of Teachers of Mathematics (ATM) held at Nottingham University, Easter 1992, Her Majesty's Staff Inspector for Mathematics appealed to the next Secretary of State for Education to grant just such a period of calm.

However, one must ask whether it is actually possible to have such a period of calm without the real risk of stagnation. Curriculum development in mathematics has, until the introduction of the National Curriculum, been gradual and consensual. As developments have taken place, new methods advocated, new topics advanced for inclusion in the curriculum, there has been a sense in which teachers have had the time to adjust, to think them through and make up their own minds about their position on the issues. The curriculum was different in different areas of the country, in different classrooms, in the hands of different teachers, and yet there was a great deal in common. It is now a part of the law of the land and so is nominally the same throughout the country. It cannot be changed or developed by individuals through experiment; what is written must be provided as a priority. If it stays the same for long periods of time it will become out of date very rapidly and further change will then have to be radical. It seems that we may have reached a situation in which the development of the curriculum goes 'underground' until sufficient pressure is exerted to force the consideration of a change. Even then that change may not come about. The curriculum is now firmly in the grasp of Parliament, and the mechanism for change or development of the curriculum in such circumstances is not at all clear. Educational arguments are not automatically political arguments. At the most absurd level, the inclusion or exclusion of any particular topic is now a political decision.

One must hope that the freedom to choose methods of teaching is kept for the teacher, otherwise all innovation will flounder, and the future will see long periods of stagnation interspersed with sharp discontinuities.

REFERENCES

Ahmed, A. (1987) *Better Mathematics. A Curriculum Development Study*. London: HMSO.
Alexander, R., Rose, J. and Woodhead, C. (1992) *Curriculum Organisation and Classroom Practice in Primary Schools. A Discussion Paper*. London: DES.
APU (1980a) *Mathematical Development: Primary Survey Report No. 1*. London: HMSO.
APU (1980b) *Mathematical Development: Secondary Survey Report No. 1*. London: HMSO.
APU (1981) *Mathematical Development: Primary Survey Report No. 2*. London: HMSO.
APU (1986) *A Review of Monitoring in Mathematics 1978–1982*. London: DES.
Cockcroft, W. H. (1982) *Mathematics Counts*. London: HMSO.
Cornforth, K. (1992) Continuity of curriculum and assessment at Key Stages 3 and 4. Talk given to the Yorkshire Branch of the Mathematical Association, at the University of Leeds, Leeds, 10 November 1992.
Davis, P. J. and Hersh, R. (1983) *The Mathematical Experience*. Harmondsworth: Penguin.
Davis, P. J. and Hersh, R. (1988) *Descartes' Dream*. Harmondsworth: Penguin.
Denvir, B., Brown, M. and Eve, P. (1987) *Attainment Targets and Assessment in the Primary Phase: Mathematics Feasibility Study*. London: DES.
DES (1987) *Report of the Task Group on Assessment and Testing*. London: DES.
DES (1991) *Mathematics in the National Curriculum (1991)*. London: HMSO.
GAIM (1988) *Graded Assessment in Mathematics Development Pack*. London: Macmillan.
Hamilton, D. (1976) *Curriculum Evaluation*. London: Open Books.
Harrison, P. (1992) SMP computer survey results. *SMP 11–16 User* **14**.
Hart, K. M. (ed.) (1981) *Children's Understanding of Mathematics: 11–16*. London: John Murray.
HMI (1991) *The Implementation of the Curricular Requirements of the Education Reform Act. Mathematics, Key Stages 1 and 3. A Report by H. M. Inspectorate on the First Year, 1989–90*. London: HMSO.
HMI (1992) *The Implementation of the Curricular Requirements of the Education Reform Act. Mathematics, Key Stages 1, 2 and 3. A Report by H. M. Inspectorate on the Second Year, 1990–91*. London: HMSO.
Howson, A. G., Keitel, C. and Kilpatrick, J. (1981) *Curriculum Development in Mathematics*. Cambridge: Cambridge University Press.
Kinder, K., Harland, J. and Wootten, M. (1991) *The Impact of School-Focused INSET on Classroom Practice*. Slough: NFER.
Mathematical Association (1989) *Mathematics Homework on a Micro*. Leicester: The Mathematical Association.
NCC (1991a) *Mathematics Programmes of Study; INSET for Key Stages 1 and 2*. York: NCC.
NCC (1991b) *Mathematics Programmes of Study; INSET for Key Stages 3 and 4*. York: NCC.
NCC (1992a) *Using and Applying Mathematics Book A. Notes for Teachers at Key Stages 1–4*. York: NCC.
NCC (1992b) *Using and Applying Mathematics Book B. INSET Handbook for Key Stages 1–4*. York: NCC.
NCC (1993) *The National Curriculum at Key Stages 1 and 2. Advice to the Secretary of State for Education, January 1993*. York: NCC.
Noss, R., Goldstein, H. and Hoyles, C. (1989) Graded assessment and learning hierarchies in mathematics. *British Educational Research Journal* **15**(2), 109–20.
Piaget, J. (1973) Comments on mathematical education. In A. G. Howson (ed.), *Developments*

in Mathematical Education. Cambridge: Cambridge University Press.

SEAC (1992a) *KS3 Mathematics. Pupils' Work Assessed*. London: SEAC.

SEAC (1992b) *KS3 School Assessment Folder (Part Three). Materials to Support the Assessment of Ma1: Using and Applying Mathematics*. London: SEAC.

Stow, M. with Foxman, D. (1988) *Mathematics Coordination; A Study of Practice in Primary and Middle Schools*. Windsor: NFER-Nelson.

Thom, R. (1973) Modern mathematics: does it exist? In A. G. Howson (ed.), *Developments in Mathematical Education*. Cambridge: Cambridge University Press.

Thwaites, B. (1972) *The School Mathematics Project. The First Ten Years*. Cambridge: Cambridge University Press.

Chapter 5

New Topics in the Mathematics Curriculum: Discrete Mathematics

John Monaghan and Anthony Orton

WHAT IS DISCRETE MATHEMATICS?

As teachers of mathematics will know, there is a distinction between discrete and continuous mathematics. Discrete mathematics deals with countable sets such as the natural numbers and the rationals, while continuous mathematics is concerned with uncountable sets, for example the real numbers. Discrete mathematics 'deals with discrete objects and finite processes' (Hart, 1991). Until relatively recently the term 'discrete', as a regular part of the language of the classroom, was usually reserved for topics within statistics. With the rise in importance of the computer in mathematics, however, the term 'discrete mathematics' has taken on a new meaning which will be explored in this chapter. According to the National Council of Teachers of Mathematics (NCTM, 1989):

> As we move toward the twenty-first century, information and its communication have become at least as important as the production of material goods. Whereas the physical or material world is most often modeled by continuous mathematics . . . the nonmaterial world of information processing requires the use of discrete (discontinuous) mathematics.

Discrete mathematics may thus be defined as the mathematics of information processing. But information processing in a mathematical sense starts with counting, so the very beginnings of discrete mathematics could certainly be considered to be when children first meet numbers in the context of counting, with which they eventually become relatively comfortable. Subsequently, and over a period of time which might be quite long, our pupils are introduced to the idea that numbers include more than just these 'natural' numbers, because fractions or rational numbers, and also integers, are regarded as numbers by their teachers. Eventually, as children progress through school, they may develop some understanding of irrationals and real numbers, and a few will also learn about complex numbers. At first, therefore, there is built-in discreteness in the understanding of numbers which develops, and indeed, in this sense, discrete mathematics has always been the major element of school mathematics, but ultimately we do wish to introduce ideas of continuity as well. It may be that, for many pupils, the grasp of real numbers by the time they leave school is very tenuous (Monaghan, 1988). In other words, our success in introducing ideas of continuity and the distinction between discreteness and continuity in connection with numbers is quite likely to be very limited, as far as most pupils are concerned. Despite clear manifestations of continuity all around us, such as in the passage of time and in growth, the journey from the discrete to the continuous within mathematics does not seem to be an easy one for many children and students. Thus discrete mathematics is sometimes claimed to be intrinsically simpler than continuous mathematics. According to the Joint Matriculation Board (JMB, 1990), 'Many consider discrete mathematics to be more accessible to "non-mathematicians".' Perhaps the earliest clear indication to pupils that mathematicians wish to emphasize a distinction between discreteness and continuity is when dealing with probability distributions. It is interesting and significant to note here, however, that discrete methods are often used to deal with continuous probability distributions.

It is not only in the pupils we teach that we observe that the passage from discrete to continuous is not easy. Throughout the recorded history of the development of mathematics, continuity has caused difficulties. Discreteness seems to have been less of a problem in a conceptual sense, though it has certainly provided difficulties in terms of manipulating numbers with ease and with accuracy. Since the time of Newton, however, many of the new developments in mathematics have revolved around what was at the time the new-found ability to deal with continuity, principally with the aid of calculus. Now we find that the computer has regenerated work in discrete mathematics, because machines can manipulate the numbers for us, and some would say this has turned the direction of the development of mathematics round. For example, we are now able to use computers to carry out integration numerically and with ease. In addition, technology demands discreteness; witness the discreteness of graph drawing on graphic calculators and on computers. Thus the computer both facilitates and demands discrete approaches to continuous situations and allows old problems such as those which could not be solved until the invention of calculus to be solved by discrete methods.

Although discrete mathematics, even at the school level, is in a sense therefore not entirely new, the idea of offering for study a subject entitled 'discrete mathematics' *is*, and any particular discrete mathematics syllabus is likely to contain some topics which are unfamiliar. Within the age range 5–16, elements of the existing mathematics curriculum can be classed as discrete mathematics, and these elements would be familiar to us all – for example, sets. For older students, however, a major purpose in defining a separate discrete mathematics syllabus, which would inevitably contain some topics not previously included in mathematics curricula – for example, graph theory – is to allow certain pupils the opportunity to focus on particular parts of mathematics which might be of great value to them and with which they can cope, when they might find other parts more difficult and less useful. Discrete mathematics has its extreme enthusiasts, inevitably, but there is as yet no indication that discrete mathematics is the new mathematics which is intended to replace the old completely for all students. There is one radical suggestion, however, which is that some students of advanced mathematics should study no calculus at all but should study discrete mathematics instead, and this will be referred to again later.

DISCRETE MATHEMATICS IN SCHOOL

If discrete mathematics is with us already, in our school curriculum at a variety of levels, it would seem appropriate to provide some examples. The curriculum for the early years of schooling offers many opportunities for investigatory approaches involving sorting and counting, and this is an example of combinatorics, which is certainly part of discrete mathematics. The example called 'Faces at Windows' is one of many investigations devised for young children by Len Frobisher. The local evidence we have from using such tasks indicates that considerable interest is aroused in children by such problems in elementary discrete mathematics, and that great success can be achieved. Examples of mathematical thinking taking place are provided by follow-up questions from the children such as 'What if they were identical twins?' and 'What if the windows were side by side like in a bus?' There are very positive educational aspects to such problems too, for they allow children to practise a number of process skills like recording, enumerating and hypothesizing. Indeed, by virtue of the small amount of content needed to support such problems, they also allow children of a wide range of ability to attempt them at their own level.

In secondary-school mathematics the topic of networks has been with us for many years (see for example SMP, 1969), with the Königsberg Bridge Problem ubiquitous as a motivator in classrooms all over the country. Indeed, networks sometimes form a basis for investigatory and course work, for example in 'Problems with Patterns and

FACES AT WINDOWS

You have two faces, one boy and one girl.

      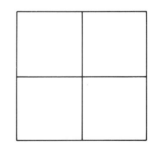

One window is made up of four small panes.

There is only room for one child to look out of each pane.

Let your two faces look out of two of the panes and record your results on squared paper.

How many different ways can you find?

Explore different ways of using the information.

Investigate other windows and families

Numbers' (JMB/Shell Centre, 1984). An example such as the one headed 'Networks' now falls within the domain of discrete mathematics.

NETWORKS

A network is a set of lines (arcs), junctions (nodes) and spaces (regions) which compose a shape. The network shown has 7 arcs, 4 nodes and 5 regions.

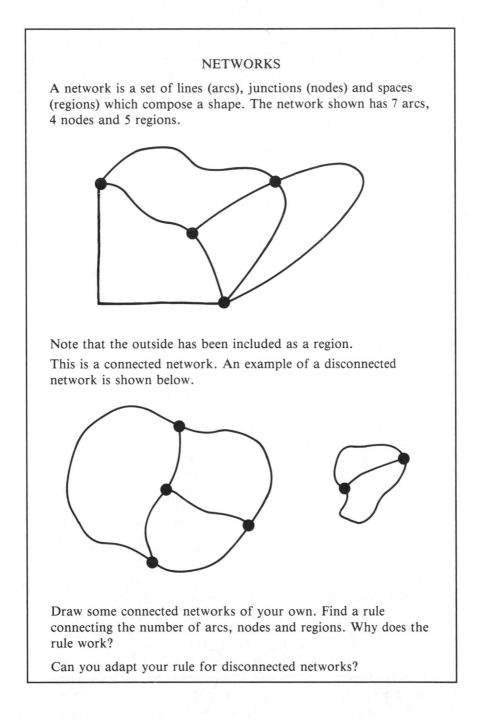

Note that the outside has been included as a region.

This is a connected network. An example of a disconnected network is shown below.

Draw some connected networks of your own. Find a rule connecting the number of arcs, nodes and regions. Why does the rule work?

Can you adapt your rule for disconnected networks?

Mathematical induction has had a place in many 16–19 curricula for a long time. The examples headed 'Induction' will be familiar to many readers. Although university mathematicians would be likely to applaud the fact that here, and all too rarely in contemporary school curricula, a mathematical proof is required, students have never found induction easy. Often the procedure can be learned almost instrumentally (see Chapter 3), but the underlying feeling that the procedure is a cheat and nothing has been proved remains. As a legitimate element of 16–19 discrete mathematics, it would seem that induction is unlikely to be particularly 'accessible to non-mathematicians'.

INDUCTION

1. Prove that the sum of the squares of the first n natural numbers is $n(n+1)(2n+1)/6$.

2. If n is a positive integer, prove that the number of subsets of $\{1, 2, 3, \ldots, n\}$ is 2^n.

COMPONENTS OF DISCRETE MATHEMATICS

From the examples provided so far it can be seen that induction, combinatorics and networks are all considered to be components of what is currently regarded as discrete mathematics, and that at an elementary level most of us already have some acquaintance with what is involved. Simple network problems at the school level are particularly important because they provide a way in to graph theory, which forms a large element of many discrete mathematics curricula. Althoen *et al.* (1991), writing about graph theory in elementary discrete mathematics, have claimed that 'It is worthwhile to study graphs at every level of the curriculum', and that graphs are particularly valuable because of the part they play in 'creating models'. Graph theory is, indeed, usually an option in the final year of undergraduate mathematical studies. Networks are of significance in the real world today through problems concerned with, for example, transportation, telephone systems and job scheduling, and the mathematical analysis of network problems typically involves developing or applying optimization algorithms for shortest paths. Here is an indication of why discrete mathematics is considered so vital to modern-day life. Dossey (1991) summed up both its importance and its inevitable association with computing in saying, 'The branch of mathematics known as discrete mathematics has rapidly grown in prominence in the past decade . . . due in large part to the many applications of its principles in business and to its close ties to computer science.' In fact, one of the justifications of discrete mathematics as a school subject is that it does enable the study of various kinds of realistic and relevant problems which are not encountered in more established curricula.

Another familiar branch of mathematics claimed by discrete mathematics is set theory. The study of sets was, of course, included within the new mathematics curricula proposed in the 1960s, but interest in sets, particularly in secondary school mathematics, has waned since then. The basic intentions of studying sets as a foundation for mathematics and as an element of the early number education of children were always relevant, but the introduction of excessive symbolism too early was misguided and there seemed to be no sensible way to test knowledge of sets in the secondary school, where testing all too often drives the curriculum. Now we are finding a reawakening of interest in sets as a component of discrete mathematics, leading both to logic and to Boolean algebra. Logic, in fact, relies heavily on induction over well-founded sets for proofs of its theorems, thus confirming the importance of the method of induction. Of course, sets were not only included, they formed the foundation for a considerable part of the new mathematics of the 1960s, for example in defining regions of the plane, thus leading to linear programming. One very early reference to this latter topic was in connection with the curriculum for 12-year-olds (Mansfield and Thompson, 1963), and it proved to be difficult for many pupils of that age. At the school level, linear programming has normally involved only two variables and problems have been devised so as to be solvable by drawing a few simple graphs. Linear programming is now a widely used method for solving particular kinds of real-life problems, and certainly fits within the discrete mathematics curriculum. Techniques such as the simplex method together with the power of modern computers now enable much more complicated problems to be solved.

Algorithms form a significant part of what has traditionally been taught in mathematics lessons. The value of learning a large number of pencil-and-paper algorithms is debatable at the moment, not least because of the existence of computers and calculators (see Chapter 12). However, the development of computing has created a need to study algorithms themselves more deeply. For example, are there a number of different algorithmic methods for solving a particular problem, and if so how do they compare in computational time needed and computational space needed? In terms of real-life applications, sorting is a regular feature of company transactions; problems such as job scheduling, equipment utilization, forecasting, stock control, budgeting and credit analysis are of daily concern. The mathematical analysis of sorting algorithms is clearly very important. Algorithms are also required in the study of networks, as we have already seen. Discrete mathematics, in fact, both incorporates the study of algorithms and also contains many algorithms in its own subject matter. Gardiner (1991) is one who has expressed concern about the algorithmic content of discrete mathematics, because 'School mathematics all too easily degenerates into a succession of meaningless routines' and 'discrete mathematics has characteristics that make it vulnerable to such degeneration'.

The relationship between discrete mathematics and computing is inevitably a close one. Discrete mathematical topics often depend on computing power for their solution and, on the other hand, there are other discrete mathematical topics which underpin the computing methods used. The relationship is therefore two-way. Another component of discrete mathematics with close ties with computing is the study of programming languages and, indeed, the study of languages generally. Programming languages are simply formal languages which are designed with specific algorithmic processes and input/output features in mind. The differences between well-known programming languages, the arguments which rage about which is best, particularly for children and other beginners, and the various advantages and disadvantages which any particular language offers should be enough to confirm that there is an important area of study here. Other aspects of discrete mathematics which relate directly to computing and which we who work in schools would probably recognize include the study of data structures such as stacks, queues, lists and trees, the study of logic gates (AND, OR, NAND, NOR), and the study of number systems such as binary, octal and hexadecimal.

The various discrete mathematics curricula devised up to now for the 16–19 age range show many differences and this has led to the criticism that there really is not a clearly defined and distinct subject here at all. Indeed, some curricula which appear to include many of the above topics are described by other names, such as 'decision mathematics' and 'algorithms and discrete models'. The wide variety of other topics which appear in them includes matrices, recursion, iteration, critical path analysis, descriptive statistics, probability and simulation, finite mathematics, decision making, chaos and game theory. A number of organizations such as the Mathematics in Education and Industry (MEI) Schools Project, as well as the various Examination Boards such as the Joint Matriculation Board (JMB) and the University of Oxford Delegacy of Local Examinations (UODLE), are producing or have already produced syllabuses in discrete mathematics (MEI Schools Project, 1990; JMB, 1990; UODLE, 1987). The next few years seem likely to provide considerable interest as we compare these syllabuses and debate their merits and demerits. At the time of writing, the dangers of degeneration into a succession of routines warned of by Gardiner are clear in some proposals, with their emphasis on learning to use a few basic network algorithms.

FOR AND AGAINST DISCRETE MATHEMATICS

It should have become clear that discrete mathematics is now a very important tool in real-life problem solving. Many of the themes noted earlier, such as networks, sorting and linear programming, are the daily mathematics of the world outside the classroom. A view often heard at the moment is that discrete mathematics will become even more important as more people see the need to use computers or are required to use them in their work. The first reason for teaching mathematics which most people give when asked is that it is useful (see Chapter 1), so it should not be surprising that mathematics educators are having to look hard at their courses, and are having to provide appropriate discrete topics. The advocates of discrete mathematics are often at pains to point out that much of the subject is more readily accessible to students than some more traditional topics, a point introduced earlier through the quotation from the JMB (1990). An example of this is in networks where, it is claimed, comparatively little theory is required before real-life applications can be tackled. Compare this with calculus, where a great deal of theory is taught, and realistic, meaningful and understandable uses can be less easy to find. Indeed, discrete mathematics has been suggested as the alternative to calculus for certain students (French, 1989), namely those whose other subjects of study are not the traditional sciences. The mathematical needs of computer science, economics, business studies, social studies, geography and even biology are, arguably, mainly provided for by discrete mathematics and not by calculus.

Another advantage of discrete mathematics arises because of the widely held view that we all need to develop and improve our skills in the use and applications of information technology (IT), because of their impact on our lives, and not only at our places of work. This demand to be able to handle applications of IT can only continue to become more pervasive in modern living, even though the extent to which individuals require information technology skills and discrete mathematics will vary. Some of the applications provide us with alternative ways of doing the same thing as before, but with what many would regard as greater ease. Spreadsheets, for example, have many applications which have not yet been fully harnessed at school level, and a simple illustration lies in maxima and minima problems. Older, alternative methods of solution would be likely to require considerable expertise and understanding within both algebra and calculus.

The arguments for and against broadening the 16–19 curriculum continue, with the current consensus view from education being that the curriculum in Britain is often too narrow and specialized (DES, 1988). Any broadening would require any one subject, like mathematics, to take up less time, and under these circumstances very careful thought about what mathematics should be taught would need to take place. Coupled with this, it is necessary in most countries to attract more students to stay in full-time education for much longer. This seems to imply that we need mathematics courses which lead to realistic and accessible applications without the need for the sustained development of theory. To some extent, and for some students, we also need easier mathematics. It should be clear that this argument is being used to support discrete mathematics as the most appropriate mathematics to satisfy the needs of many students within a broader 16–19 curriculum. There is possibly also a bonus in that much discrete mathematics, though by no means all, is of fairly recent origin, and this can encourage students to think of mathematics as a subject which is live and still evolving. Too often students regard mathematics as a body of knowledge whose laws were set in stone centuries ago.

Discrete mathematics cannot be introduced into a mathematics curriculum for all without making space for it, so what goes out? Given past experience, it is not going to be easy to be radical, for the tendency has always been to add without taking away. The problem is exacerbated by the fact that there is also pressure for space to be made to develop process and 'core' skills (see Chapter 1 and NCC, 1990) in our pupils and students. Broadening the whole curriculum leaves less time for mathematics; course work and problem-solving activities within mathematics eat up valuable time. Something substantial will have to go, and the target is calculus. There are those who mourn the passing of Euclidean geometry, even though the majority of pupils could not do it; are we to have similar withdrawal problems with calculus? It is argued, admittedly by the firm advocates of discrete mathematics, that calculus is increasingly redundant in the age of new technology. It could also be said that calculus, like Euclidean geometry for 11–16-year-olds, is too difficult for most 16–19-year-old students, certainly if one hopes for more than just instrumental understanding (see Chapter 3). It was different when only a few very bright students stayed in education beyond 16+, but students of a much wider range of accomplishment are in our 16–19 classrooms nowadays.

Removing calculus from the curriculum, however, would seem to have far more serious implications for the rest of mathematics than rejecting Euclidean geometry. Calculus still forms the core of first-year studies in many university mathematics departments, and is still essential for many aspects of the pure sciences. Mechanics, which is strongly defended as a component of 16–19 mathematics by university applied mathematicians, still makes use of calculus, and not all statistics is discrete. There is also the danger that teachers and curriculum developers at upper-school and university levels would still continue to assume considerable knowledge of calculus when it might not be there, just as we all too easily assume a knowledge of geometry which is now unreasonable, for example when teaching coordinate geometry. What is more, removing calculus from the curriculum would inevitably necessitate a very thorough examination of all the mathematics which comes before, namely what is described in the USA, with great significance, as 'pre-calculus'. To some people, the only solution is what has already been mentioned, that is, separate courses or modules which allow students choice according to their interests and future requirements. A module in calculus could be studied by those whose interests lay either in mathematics alone or in the physical sciences, while an alternative module in discrete mathematics could be studied by other students. There is then a danger of our educating two classes of mathematics students, an upper class for whom difficult mathematics has been included within the curriculum, and a lower class for whom calculus has been omitted and discrete mathematics included instead. Such alternative courses also appear to complicate discussions further about whether all students should follow a common core of mathematics in the 16–19 age range, and, if so, what that core should be. In fact, there is an argument which says that the emergence of discrete mathematics as an alternative renders the objective of agreeing a common core unattainable (French, 1989).

There is some criticism of discrete mathematics in comparison with the mathematics of the traditional, established curriculum. The argument is that a discrete mathematics curriculum, because of the nature of what we understand the subject to be, is inevitably merely a framework within which to fit a number of separate (dare one say discrete?) topics which do not gel into a coherent whole in the same way as other mathematics modules would be likely to do. Indeed, this view appears to be supported by the differences in content between discrete mathematics modules so far proposed in Britain. Whether such lack of coherence matters seems open to individual interpretation. Another possible criticism is that some discrete mathematics is already being taught within existing curricula and much of the rest is to do with using computers, so why should it not be part of computer studies rather than mathematics? Furthermore, some discrete mathematics curricula appear to consist largely of the application of a few standard network algorithms. The examination board syllabuses referred to earlier could be seen, in the worst possible light, as examples of an interpretation of discrete mathematics as merely comprising a limited number of network algorithms. Discrete mathematics must clearly go beyond such crude characterizations. Also, topics in discrete mathematics can be transient, it seems. Some years ago we were all being exhorted to teach flowcharting within mathematics, but that encouragement now seems to have disappeared; indeed, at the moment computer science seems positively to discourage it.

The last major revolution in the mathematics curriculum took place in the 1960s and 1970s and is sometimes known as the 'modern mathematics' movement. Many mistakes were made at that time, though we must not overlook the fact that the curriculum for all of us has never been the same since, and that some topics have certainly stood the test of time; for example, probability, statistics and transformations in geometry. Overall, our pupils are better educated for current needs than they would be had no changes taken place since the 1950s, but we do need to regard that earlier revolution as a cautionary tale. In an ideal world we should be able to avoid the kind of yo-yo effect of bringing in topics, then abandoning them, and then bringing them back again, as may well happen with sets. Discrete mathematics is not going to go away, but decisions need to be taken with care. Piecemeal curriculum change could create just as many problems as the wholesale changes proposed in the 1960s.

In conclusion, it should not be the case that discrete and continuous mathematics are regarded as alternatives. Both aspects have complemented each other for centuries, and this should be reflected when we teach the subject called mathematics. It is natural that, as discrete mathematics has increased in importance, it should be discussed as a separate entity, but as it becomes established we need once again to develop unified approaches to mathematics which suit our pupils and students. Seidman and Rice (1986) have indicated how this can be done at the upper secondary to university interface. The challenge for curriculum planners and developers is to do the same at all levels of the curriculum.

REFERENCES

Althoen, S. C., Brown, J. L. and Bumcrot, R. J. (1991) Graph chasing across the curriculum: paths, circuits, and applications. In M. J. Kenney and C. R. Hirsch (eds), *Discrete Mathematics across the Curriculum, K-12*. Reston, VA: NCTM.

DES (1988) *Advancing A Levels (The Higginson Report)*. London: HMSO.

Dossey, J. A. (1991) Discrete mathematics: the math for our time. In M. J. Kenney and C. R. Hirsch (eds), *Discrete Mathematics across the Curriculum, K-12*. Reston, VA: NCTM.

French, S. (1989) The common core for mathematics: do we need one? *Teaching Mathematics and its Applications* 8(4), 180-3.

Gardiner, A. D. (1991) A cautionary note. In M. J. Kenney and C. R. Hirsch (eds), *Discrete Mathematics across the Curriculum, K-12*. Reston, VA: NCTM.

Hart, E. W. (1991) Discrete mathematics: an exciting and necessary addition to the secondary school curriculum. In M. J. Kenney and C. R. Hirsch (eds), *Discrete Mathematics across the Curriculum, K-12*. Reston, VA: NCTM.

JMB (1990) *Mathematics (Algorithms and Discrete Models)*. Manchester: Joint Matriculation Board.

JMB/Shell Centre for Mathematical Education (1984) *Problems with Patterns and Numbers*. Nottingham: University of Nottingham Shell Centre for Mathematical Education.

Mansfield, D. E. and Thompson, D. (1963) *Mathematics: A New Approach, Book 2*. London: Chatto and Windus.

MEI Schools Project (1990) *Structured Sixth Form Mathematics Scheme*. Bath: The Mathematics in Education and Industry Schools Project.

Monaghan, J. (1988) Real mathematics. *The Mathematical Gazette* **72**, 276-81.

NCC (1990) *Core Skills 16-19: A Response to the Secretary of State*. York: National Curriculum Council.

NCTM (1989) *Curriculum and Evaluation Standards for School Mathematics*. Reston, VA: NCTM.

Seidman, S. and Rice, M. (1986) A fundamental course in higher mathematics incorporating discrete and continuous themes. In A. G. Howson and J.-P. Kahane (eds), *The Influence of Computers and Informatics on Mathematics and its Teaching*. Cambridge: Cambridge University Press.

SMP (1969) *Teacher's Guide for Book B*. Cambridge: Cambridge University Press.

UODLE (1987) *Decision Mathematics Advanced Supplementary Level*. Oxford: University of Oxford Delegacy of Local Examinations.

Chapter 6

New Topics in the Mathematics Curriculum: Fractal Mathematics

Dave Carter

INTRODUCTION

Traditionally, school geometry has focused on the study of the properties of two- and three-dimensional shapes – the former bounded by straight lines or smooth curves, the latter by planes or smoothly curved surfaces. Although we see many examples in the environment, these are mostly artificial. Crystals, admittedly, are examples of three-dimensional shapes that are bounded by planes and that, although they can be manufactured in the laboratory, do occur naturally, having been formed under intense heat and pressure within the earth's crust (Bronowski, 1973). Most natural shapes, however, tend to be crinkly and appear to be irregular. Behind this apparent irregularity, though, there is often surprising regularity. As the Mathematical Association (1991) observed:

> A cauliflower head can be broken down into branches – each of which is similar to the whole head; each branch can be further divided into sub-branches and the process continued until the individual florets are isolated. At each stage the shape of the unit is similar to the whole head.

Mandelbrot (1977) first coined the word 'fractal' in 1975 to describe a geometric shape that is self-similar, having infinite detail at all scales. He further explains that a shape is self-similar when each piece of the shape is geometrically similar to the whole. A simple example of a fractal is the familiar snowflake curve, which is built up by starting with an equilateral triangle and replacing each side according to the rule shown in Figure 6.1. At each subsequent stage every straight line segment is replaced according to the same rule. The process is continued indefinitely. The original triangle and the first two stages of the process are shown in Figure 6.2.

Figure 6.1

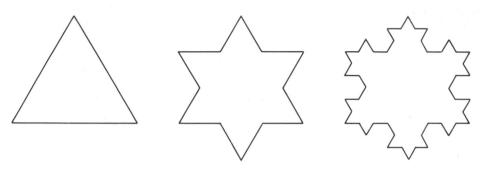

Figure 6.2

The familiar dragon curve, the subject of the video *Dragonfold* (available from Edward Patterson Associates Ltd), and the fractal depicted on the Leapfrogs 'Fractal' poster (available from Tarquin Publications), are other simple examples of fractals. The dragon curve is generated by starting with a straight line segment (Figure 6.3a). This straight line can be regarded as the hypotenuse of an isosceles right-angled triangle. The replacement rule is to replace the hypotenuse by the two shorter sides (Figure 6.3b). Subsequently the replacements are placed on alternate sides of succeeding straight lines (Figure 6.3c). Figure 6.4 shows the result of the eleventh iteration of the process – notice that the figure appears to be made up of blocks which are similar. If the process could be continued indefinitely, the dragon curve would indeed be self-similar; enlargements of it would show infinite detail at all scales.

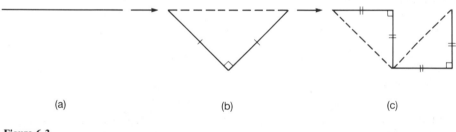

(a)   (b)   (c)

Figure 6.3

In general the generation of better and better approximations to simple fractals is by iteration – a feedback process. At any stage each straight line segment is replaced by a combination of straight line segments according to some rule, and the newly formed straight line segments are themselves replaced at the next stage by the same rule. The drawing of these simple fractals is an activity suitable for 12–13-year-old pupils; such an activity can be extended to the calculation of the length of the curve at each stage, and, in the cases where the original figure is a closed polygon, to the calculation of the area enclosed at each stage. Such processes are easy to set up on a microcomputer using recursion techniques, and can provide pupils/students with a visual representation of sequences which diverge (the perimeters at successive stages) and sequences which converge (the areas inside closed polygons at successive stages). The presence of the

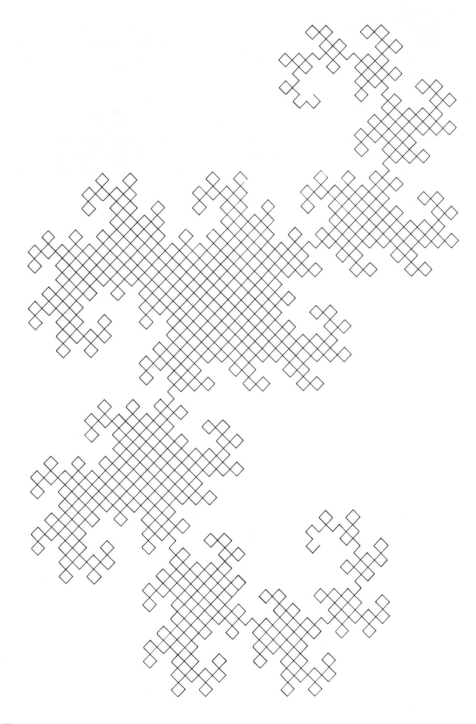

Figure 6.4

microcomputer in the mathematics classroom demands a rethink of the way we approach many topics. What is suggested here makes it possible to use these activities as a basis for discussing the concepts of infinity and of limit at an intuitive level.

FRACTAL MATHEMATICS IN SCHOOL

Some fractals tessellate and these can be used as the basis of some interesting investigational work (Orton, 1991). It is clear that the geometry of fractals can lead to the study of number sequences which can be generalized as algebraic formulae in the form of recurrence relations, a topic included in many school syllabuses. Other examples of feedback processes are growth laws, many of which appear to be well modelled by the linear transformation

$$x \to (1 + r)x$$

where r is a fixed rate. This transformation can be expressed as the recurrence relation

$$x_{n+1} = (1 + r)x_n.$$

An everyday example of this relation is the growth of money invested at a fixed rate, r per cent, compound interest. However, in reality, growth is not without limit – particularly in ecological systems. As a population reaches its limit, the rate of growth often decreases steadily to zero, which suggests that r should be replaced by by $r - bx_n$. This means that

$$x_{n+1} = (1 + r - bx_n)x_n$$

may be a better model in certain circumstances. This relation is non-linear; the term bx_n^2 has a remarkable effect on the behaviour of the sequences of numbers generated. Peitgen and Saupe (1988) discuss the behaviour of the associated recurrence relation

$$P_{n+1} = kP_n(1 - P_n).$$

and show that population growth, particularly in the case of two competing populations, depends not only on the initial conditions but more importantly on the value of k, which might be interpreted as a measure of the environmental conditions.

Such non-linear transformations have interested mathematicians, physical scientists, engineers and the medical profession in recent years, since they provide a means of analysing phenomena such as optical resonance, chemical reactions, turbulence and heart rhythms which were not susceptible to analysis using linear methods (see Hall, 1991, and many articles which have appeared in *Nature* in recent years).

Iteration has found its way into school syllabuses. In the National Curriculum for England and Wales, for instance, one example given is 'Solve quadratic equations by using factors, the common formula, completing the square or iteration as appropriate'. Clearly the inclusion of iteration is to prepare the very ablest pupils for situations where there is no other suitable method; for example, the solution of f(x) = 0 using Newton's method, which occurs later on in mathematics syllabuses.

To solve, for example,

$$x^2 - x + c = 0$$

(*c* being an arbitrary real constant) by iteration, one can construct six different recurrence relations, some of which produce rapidly convergent sequences of approximations to one of the roots. One of these six relations,

$$x_{n+1} = x_n^2 + c,$$

is not an obvious choice. However, it is worthy of exploration because it reveals a great deal about the dynamics of the 'squaring and adding' process. Clearly the success of the iterative process depends on an appropriate choice of starting value x_0. However, for values of c outside the range $-2 \leqslant c \leqslant 0.25$, no matter what value of x_0 is taken the iteration will not work (the sequence of approximations is divergent). Even if c lies within this range and a suitable starting value x_0 is chosen, the resulting sequence of approximations to one of the roots may behave strangely.

For $-0.75 < c \leqslant +0.25$	the sequence converges to the smaller of the two roots.
For $-1.25 < c \leqslant -0.75$	the sequence at first converges but then cycles between two values.
For $-1.4 < c \leqslant -1.25$	'bifurcation' continues and produces cycles of 4, 8, 16 . . . values.
For $-2 \leqslant c \leqslant -1.4$	the behaviour of the sequence of approximations is very erratic indeed; for example, for $c = -1.755$, the sequence eventually cycles between three values, and for $c = -1.99$, different values of x_0 produce different sequences of which none converges but all are bounded. Chaos!

These behaviours are well illustrated by the use of a graph-drawing package on a microcomputer. The graphs of $y = x^2 + c$ and $y = x$ are plotted and by the use of a suitable drawing procedure the path of each iteration can be illustrated, showing clearly when the process converges to a limit, cycles or behaves chaotically. Figure 6.5 shows graphically the behaviour of the process for (a) $c = -0.6$, (b) $c = -1.0$, and (c) $c = -1.9$.

An interesting variation on the theme of repeated squaring is in the exploration of sequences of numbers generated by the mappings

$$x \to x^2 \text{ modulo } 10, 100.$$

Using calculators, 12–13-year-old pupils have been fascinated by the rich variety of number patterns they were challenged to find when these problems were set in the form of a competition. Indeed, the general mapping

$$x \to x^2 + c \quad \text{modulo } m \ (0 \leqslant c < m)$$

makes for an intriguing investigation.

The effect of the quadratic transformation,

$$z \to z^2 + (p + iq),$$

where p, q are arbitrary real constants, applied repeatedly to the complex z-plane, is even

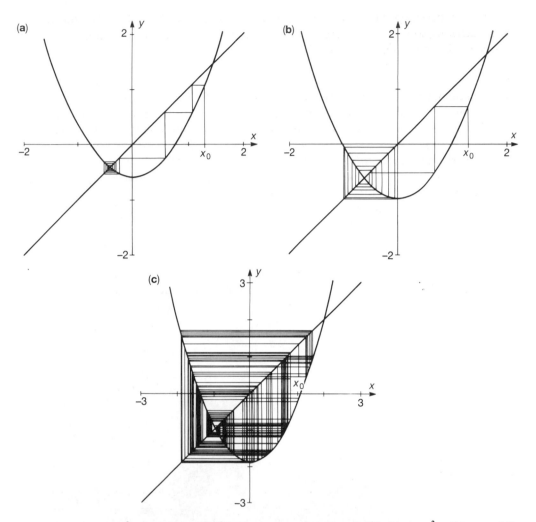

Figure 6.5 *(a)* $x = x^2 - 0.6$; $x_0 = 1.3$. *The process converges to* $x \doteq 0.422$. *(b)* $x = x^2 - 1.0$; $x_0 = 1.3$. *The process eventually cycles between the values* $x = -1$ *and* $x = 0$. *(c)* $x = x^2 - 1.9$; $x_0 = 1.3$. *The process is bounded and chaotic!*

more remarkable. Pictures, generated by microcomputer, of the effect of the mapping are very striking (see, for example, Peitgen and Richter, 1986; Peitgen and Saupe, 1988). For all values of p, q some absolute values of z grow without limit (are attracted to infinity), but for some values of p, q there are also absolute values of z which are trapped within a region of attraction (are attracted by a second attractor). For certain values of p, q this second attractor behaves chaotically. In cases where there is a second attractor, the complex plane is divided into three sets of points – those attracted to infinity (usually coloured according to the number of iterations needed for the absolute value of z to exceed an arbitrary value), those attracted to a second attractor (usually coloured black), and those on the boundary of the region of attraction. This third set is invariant under the transformation, but the iterations of these boundary values are chaotic. This boundary set is called a Julia set – after the French mathematician Gaston Julia (1893–1978).

Depending on the values of p and q, the Julia set can be continuous and linearly connected, forming a fractal curve, or a collection of disjoint sets. The set of values of p, q which give rise to a Julia set was discovered in 1980 by Benoit Mandelbrot (a research mathematician at IBM and a former student of Julia) and is called the Mandelbrot set, illustrated in Figure 6.6 (Peitgen and Richter, 1986). The boundary of the Mandelbrot set is also a fractal, and 'zooming' into the boundary reveals tendrils, whorls and spirals – spectacles of incredible beauty. Just outside the boundary lie an infinite number of copies of the set but each one different in detail from all the others. Most are invisible to the naked eye, but enlarging the area just outside the boundary reveals some of them – more and more as the magnification increases, and each one joined by filaments to the parent Mandelbrot. The number of Julia sets which can be generated by the quadratic transformation is infinite, but there is only one Mandelbrot set, which can be regarded as the directory of all values of p, q which give rise to Julia sets.

Other polynomial functions of z, or even trigonometric and exponential functions, can be used as the basis of an iterative procedure to produce spectacular Julia sets. Finding the roots of $z^n + 1 = 0$ using Newton's method produces some very interesting

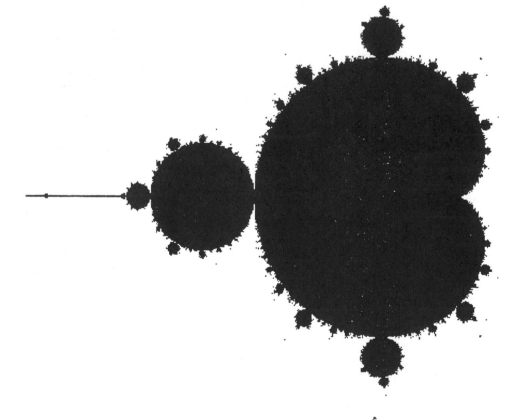

Figure 6.6

Julia sets; Figure 6.7 shows the Julia set generated by graphic iteration when locating the roots of $z^3 - 1 = 0$.

Figure 6.7

All the the examples of fractals discussed so far are constructed by the use of what are known as deterministic algorithms; that is, the outcome is determined completely by the choice of rule and an initial value.

The study of 2×2 matrices and their effect on the plane has become well established in the school mathematics curriculum over the past thirty years, although recent trends in many countries suggest that, in future, this rich area of study will only be available to those students who follow advanced school mathematics courses. Consider the four transformations represented by the following (Barnsley, 1988):

$$T_1 = \begin{pmatrix} 0.85 & 0.04 \\ -0.04 & 0.05 \end{pmatrix} + \begin{pmatrix} 0 \\ 1.6 \end{pmatrix} \qquad T_2 = \begin{pmatrix} -0.15 & 0.28 \\ 0.26 & 0.24 \end{pmatrix} + \begin{pmatrix} 0 \\ 0.44 \end{pmatrix}$$

$$T_3 = \begin{pmatrix} 0.2 & -0.26 \\ 0.23 & 0.22 \end{pmatrix} + \begin{pmatrix} 0 \\ 1.6 \end{pmatrix} \qquad T_4 = \begin{pmatrix} 0 & 0 \\ 0 & 1.6 \end{pmatrix} + \begin{pmatrix} 0 \\ 0 \end{pmatrix}$$

Taking the origin as starting point and choosing one of the above transformations at random, with probability proportional to the value of the determinant of the matrix (the determinant of the matrix in $T_4 = 0$, so choose T_4 with probability 0.001, say), plot the position of its image after applying the transformation. Then repeat the process for each of the successive images. The process is easy to set up on a microcomputer and what appears on the screen is unexpected – a computer-generated picture of a fern, shown in Figure 6.8. Barnsley (1988) discusses general algorithms for producing fractal models using contraction mappings on metric spaces. The process illustrated above is not deterministic because the paths for the iterative procedure are chosen by chance; the process is an example of what is called a random iteration algorithm.

Figure 6.8

The beauty of these computer-generated fractals has captured the imagination of the public at large – as witnessed by the use of fractal images on T-shirts, mugs, calendars, posters, postcards, greetings cards and the like. Atkins (1989) suggests that a new era has dawned, where science subsumes art. Does 'painting by numbers' constitute art? Whatever the answer, there are now many examples of fractal models of reality used as illustrations in books (Mandelbrot, 1977, 1990; Peitgen and Saupe, 1988) and as backdrops on film and television sets. Very often these fractal models are so lifelike that the eye is easily deceived. Another art form in which the use of fractal models is making an impact is music composition (Gardner, 1992).

However, the most significant recent application of fractal models is in the storage of image data for display in computer graphics, desktop publishing and multimedia systems such as CD-ROM. Data storage of images with high information content can be significantly reduced by using image compression methods. One of the most powerful of these is the fractal transform invented by Michael Barnsley in 1988 (Wright, 1992), which is based on affine transformations in the Euclidean plane (Barnsley, 1988).

Fractal mathematics offers exciting opportunities for the exploration of new areas of mathematics using standard procedures in a stimulating way. Sources of ideas for using these opportunities in the classroom abound, and many contain programs for computer activity; for example, Noss (1985), Bannon (1991), Frantz and Lazarnick (1991). However, the best collection of classroom activities can be found in Peitgen *et al.* (1991, 1992) supported by background material for the teacher in Peitgen, Jurgens and Saupe (1992a, 1992b). Stewart (1990) provides a good, simple introduction to chaos theory, while Lauwerier (1991) presents the mathematical basis of fractals at an appropriate level for students with a working knowledge of complex numbers.

CONCLUSION

Having read thus far, the reader may well ask 'So what is the issue?' Recent curriculum initiatives, in spite of paying lip-service to the cross-curricular aspects of mathematics, suggest a fragmented approach to the teaching of the subject. Moreover, assessment requirements emphasize knowledge learned for its own sake. There is a danger that, in our teaching, we shall fail to deliver the message that knowledge has a purpose, mathematics is essentially a problem-solving tool, and its elegance lies in its power to provide good models of the natural world. Mandelbrot (1990) concludes his article on the following optimistic note:

> A final satisfying spin-off from fractal pictures is that their attractiveness seems to appeal to the young and is having an influence on restoring an interest in science. Many people hope that the Mandelbrot set and other fractal pictures, now appearing on T-shirts and posters, will help to give the young a feeling for the beauty and eloquence of mathematics and its profound relationship with the real world.

One of our aims in teaching mathematics (see Chapter 1) should indeed be to 'give the young a feeling for the beauty and eloquence of mathematics and its profound relationship with the real world', and it is difficult to see how the mathematics teacher can ignore an aspect of mathematics that is already known by many young people and which is accessible with quite elementary mathematical ideas.

REFERENCES

Atkins, P. W. (1989) The rose, the lion, and the ultimate oyster. *Modern Painters* **2**, 50–5.
Bannon, T. J. (1991) Fractals and transformations. *Mathematics Teacher* **84**, 178–85.

Barnsley, M. F. (1988) *Fractals Everywhere*. San Diego: Academic Press.

Bronowski, J. (1973) The music of the spheres. *The Ascent of Man,* Chapter 5. London: The British Broadcasting Corporation.

Frantz, M. and Lazarnick, S. (1991) The Mandelbrot set in the classroom. *Mathematics Teacher* **84**, 173–7.

Gardner, M. (1992) *Fractal Music, Hypercards and More: Mathematical Recreations from Scientific American*. New York: Freeman.

Hall, N. (ed.) (1991) *The New Scientist Guide to Chaos*. Harmondsworth: Penguin.

Lauwerier, H. A. (1991) *Fractals*. Harmondsworth: Penguin.

Mandelbrot, B. (1977) *The Fractal Geometry of Nature*. New York: Freeman.

Mandelbrot, B. (1990) Fractals – a geometry of nature. *New Scientist*, 15 September, 38–43.

Mathematical Association (1991) *Fractal Calendar 1992*. Leicester: The Mathematical Association.

Noss, R. (1985) Fractals, turtles and snowflakes. *Micromath* **1**(1), 31–5.

Orton, A. (1991) From tessellations to fractals. *Mathematics in School* **20**, 30–1.

Peitgen, H.-O. and Richter, P. H. (1986) *The Beauty of Fractals*. New York: Springer-Verlag.

Peitgen, H.-O. and Saupe, D. (eds) (1988) *The Science of Fractal Images*. New York: Springer-Verlag.

Peitgen, H.-O., Jurgens, H. and Saupe, D. (1992a) *Fractals for the Classroom. Part One: Introduction to Fractals and Chaos*. New York: NCTM/Springer-Verlag.

Peitgen, H.-O., Jurgens, H. and Saupe, D. (1992b) *Fractals for the Classroom. Part Two: Complex Systems and Mandelbrot Set*. New York: NCTM/Springer-Verlag.

Peitgen, H.-O., Jurgens, H., Saupe, D., Maletsky, E., Perciante, T. and Yunker. L. (1991) *Fractals for the Classroom: Strategic Activities. Volume One*. New York: NCTM/Springer-Verlag.

Peitgen, H.-O., Jurgens, H., Saupe, D., Maletsky, E., Perciante, T. and Yunker, L. (1992) *Fractals for the Classroom: Strategic Activities. Volume Two*. New York: NCTM/Springer-Verlag.

Stewart, I. (1990) *Does God Play Dice?* Harmondsworth: Penguin.

Wright, A. (1992) FRACTALS transform image compression. *Electronics World + Wireless World*, March, 208–11.

Chapter 7

Language and Mathematics

William Gibbs and Jean Orton

THE LANGUAGE OF NUMBERS

Read these sentences aloud, listening to and reflecting on the way you say the numbers involved.

> The first man landed on the moon in 1964.
>
> The Channel Tunnel will be finished in 2002.
>
> The time is 11.00.
>
> Your new PIN number is 4604.
>
> My ticket to London cost £42.39.
>
> Could you cut a piece that is exactly 9.69 metres long?
>
> The attendance at last week's game was 12,325.
>
> To reach Brecon follow the A40 and then take the B1464.
>
> My phone number is 0874630220.
>
> The next shower of meteorites is expected to last from 2009 to 2136.

This exercise reveals not only the variety of ways in which numbers can be read but also the complexity of the conventions that unconsciously govern our use of number language. For example, the context of a number may determine whether we use an expanded or a digital form, but form may vary even within the same context. The A40 is the A forty but the B1464 is the B one four six four. We use different conventions for dates and times, for money and for phone numbers. Indeed there may be dates which we are not sure how to read as the convention has not yet been established. How did you read the last two dates in the final sentence above? There may even be individual differences and preferences for the form to be used. Some people read 0874 as zero eight seven four while others prefer to say oh eight seven four. With the increasing use of numbers which have neither cardinal nor ordinal meaning but exist solely as codes, and with the increased use of digital phones and cash points, there is a discernible change in the way numbers are being said. At a recent election, one of the returning officers gave his results in the following way:

> George William Smith, two one four six six,
>
> Mary Howard Jones, one eight two five two.

The way in which decimals are read reveals interesting variations. Feedback from a recent survey (Threlfall, 1993) suggests that many children read 7.61 as seven point sixty-one. This may be the influence of the form of language used with decimal currency. The natural everyday form of words for currency in which £7.61 is said as seven pounds sixty-one is transferred by the child to the reading of decimals in general. Such examples remind us that even in what may seem a simple area within the realm of mathematical language the conventions are complicated and liable to change.

Throughout this chapter we cannot do more than point to some of the studies of the complex interconnection between language and mathematics, chart some of the areas that have been studied in depth, and suggest questions that might lead to profitable future consideration. Where possible the issues have been related to learning mathematics through the medium of both a first and a second language.

THE MATHEMATICAL REGISTER

The special vocabulary used in mathematics, the mathematical register, was an early focus of attention. Otterburn and Nicholson (1976) investigated school children's understanding of a variety of mathematical terms and demonstrated that there were many mathematical words teachers commonly use that children were not able to explain. These results highlighted a problem of which many teachers were unaware, but there are dangers in using children's ability to explain words as an absolute measure of their understanding of technical mathematical words, an understanding which clearly may act at different levels. For example, a child may be able to respond to a word without being able to recall it, or to use it without being able to define it.

Some recent work undertaken as part of the Assessment in Key Stage 2 Project (Threlfall, 1993) studied the responses made by children aged between 7 and 11 to two questions, one requiring the children to name a shape when shown its picture, the other requiring children to identify the shape when given its name (see table).

	Percentage of children able to name the shape when shown a picture	Percentage of children able to identify the shape when given the name
Rectangle	91	96
Triangle	80	96
Cube	76	85
Hexagon	66	82
Cylinder	54	85
Sphere	52	75
Pentagon	51	76
Pyramid	47	91
Cuboid	46	83

These results suggest that it is considerably more difficult for children to recall the correct vocabulary item associated with a mathematical concept than to respond to it. For example, 91 percent responded to the word 'pyramid'. In linguistic terms they acted at the 'receptive' level, whereas only 47 percent acted at the 'productive' level by recalling the same word and using it to describe a picture. This research illustrates two of the levels, the receptive and productive, at which a mathematical term can be used within the linguistic dimension, and this reflects the knowledge we all have that we can respond to a wider vocabulary than we can use ourselves. However, items within the mathematical register are met not only within a linguistic dimension but also within cognitive and contextual dimensions. The cognitive dimension measures the level of difficulty or complexity of the mathematical concept involved and the contextual dimension the level of contextual support given. This provides a three-dimensional model with which to analyse mathematical language tasks.

Dimension	Range		
Linguistic	Receptive	⟵——⟶	Productive
Cognitive	Undemanding	⟵——⟶	Demanding
Contextual	Supported	⟵——⟶	Unsupported

This model creates eight domains with which to analyse tasks. Consider the following two tasks:

Task A: Point to the shape which is a circle.

Task B: Describe a cone.

Task A belongs to the linguistically receptive, cognitively undemanding and contextually supported domain. By contrast, Task B belongs to the linguistically productive, cognitively demanding and contextually unsupported domain. If, however, a cone was provided in Task B it would then have contextual support. Such an analysis of mathematical tasks suggests a strategy for introducing new mathematical words to children in the classroom allowing for progression from tasks lying within the easier domains to those in the harder domains.

Pimm (1987) has written clearly and with fascinating examples about how the mathematical register of a language expands over time to meet the demands of specialist language use. We must remember, for example, that English was not thought by some in the seventeenth century to be suitable for learning or teaching mathematics (Wilson, 1981), and since that time English has rapidly expanded its register to include many Latin and Greek words which are now part of the regular currency of mathematics. Such is the method by which every language expands and grows to meet the demands placed upon it, and there seems no reason why other languages such as Swahili or Kikuyu or Yoruba, which may lack certain technical words, should not grow in the same way.

But Pimm and others have also drawn our attention to areas of difficulty and confusion that can exist within the use of the mathematical register and in particular to the following potential traps.

Words have one meaning in everyday use and another in the mathematical register and can therefore be ambiguous.

Just a small sample of words in this category will make the point that many words used commonly in the mathematics lesson have a mathematical meaning as well as a common one: base, common, difference, even, improper, odd, real, right, root, similar, times, volume. Are children confused by these words or are they able to distinguish meaning due to the context within which these words are used? The context, it might be assumed, would make the meaning clear. However, the research (Durkin and Shire, 1991) does suggest that children tend to make errors of interpretation based on the common everyday use of a word, and though these errors decrease with age they still continue to occur. Working with students in years 12 and 13, Monaghan (1991) has shown that even at this level the everyday meaning of words and phrases such as 'tends to' and 'converges' can impair students' understanding of the mathematical concept of limit.

Words can change their meaning depending on their context within the mathematics lesson.

Examples of this would be the use of the word 'right' in the phrases 'right angle' and 'right side', or the use of the word 'vertical' when used by the teacher one moment to describe a vertical line in the room and the next to describe a vertical axis which may be truly vertical when drawn by the teacher on the chalkboard but horizontal when drawn by students in their books. Other words used with two different meanings in mathematics include 'tangent', 'base', 'square', 'cube', 'second' and 'third'. 'Length' and 'height' have both a general meaning when used in the mathematics classroom and also specific meanings when used in relation to rectangles and triangles. Particular attention has been drawn by Durkin and Shire to words which are used with one meaning in numerical situations and operations but with different meaning in spatial contexts. Consider the following phrases and their potential for confusion:

> A big number. A big square.
>
> A low number. A low table.
>
> The numbers 1, 2, 3, 4, 5 are going up.
>
> Think of a high number.
>
> Confusion can be caused by words that sound alike, for example 'some' and 'sum'.

A special difficulty exists in spoken English in distinguishing between the tens and the teens, for example 'sixty' and 'sixteen' sound alike. This is a problem that does not occur in many other counting systems and seems to present special difficulties for second-language learners of English (Garbe, 1985).

There are dangers associated with linking a mathematical label to only one particular type of example.

Words in mathematics can become associated with particular representations. For example, learners who have only seen examples of perpendicular lines that are drawn parallel to the edges of the book or chalkboard fail to identify examples in other orientations (Skemp, 1971).

Confusion can arise from a mathematical word used out of its context.

Messenger (1991) cites the example of a teacher who is telling the children how to find the answer to this addition:

$$\begin{array}{r} \text{m cm} \\ 3\ \ 45 \\ +\quad 1\ \ 20 \end{array}$$

The teacher encourages the children to add 'the numerator and the denominator'. Here the teacher has taken these words to mean 'the number on top' and 'the number underneath', which are common ways of explaining these terms. Many teachers use the word 'sum' in the same potentially confusing way.

Confusion can exist between 'experts' or teachers in their use of words.

A common example is the word 'billion', which now, due to inflation and the influence of America, is used almost every day on the radio or television news to mean one thousand million whereas it is still used in many mathematical texts to mean a million million. Teachers of mathematics differ in their interpretation of the words 'oblong' and 'diamond', and Kouba (1989) has shown that teachers of mathematics and science may ascribe different meanings to the same words or phrases. A science teacher, for example, may describe all graphs where the data can be represented by lines, be they curved or straight, as 'line graphs'. A mathematics teacher, on the other hand, may only use the phrase to describe straight line graphs. Yet another definition is used by Hollands (1980), who defines a line graph as 'a graph or chart where the bars of a bar chart are replaced by lines'. Kouba suggests that similar differences of interpretation exist with the words 'variables' and 'formula'. There are probably many more.

The power that certain words hold to trigger children to think in a particular way can lead to mathematical disaster.

The words 'altogether' and 'more' in a word problem prompt many children to use the operation of addition; the words 'less' and 'take away' prompt the operation of subtraction. These words are often described as 'key' or 'cue' words and in the majority of word problems they trigger the appropriate response. Consider, however, this problem, in which the triggered operation of addition is inappropriate:

> Peter has three more marbles than John.
>
> If Peter has 7 marbles then how many has John?

Stockdale (1991), in an analysis of primary-level American textbooks, has shown that the only cue words linked consistently with a single operation were 'percent' and 'equal'.

Discrepancy may exist between the meaning a teacher ascribes to a word and the meaning it carries for a child.

One of the benefits of the constructivist approach to learning is that it has once again focused attention on the dynamic nature of the meaning of words and how meaning alters as the learner meets new situations and new uses. For teachers, this raises more acutely the question of how to monitor the learner's meaning and interpretation of mathematical language and how best to develop strategies that encourage pupils to modify, expand and develop meaning.

THE MATHEMATICAL REGISTER AND SECOND-LANGUAGE LEARNING

Before leaving the topic of the mathematical register, it is interesting to pose the question of how consideration of the register can throw light on the teaching and learning of mathematics in a second language. This is a problem met by some British pupils but also by millions of children in other parts of the world. In the 1960s and 1970s a major focus of concern among mathematics educators in Africa, as shown by the reports of the seminars in Lesotho (National University of Lesotho, 1981) and Kenya (UNESCO, 1974) on mathematics and language, was the absence of an adequate mathematical register in the first language. For example, in Hausa, the language of much of Nigeria, there was no word for triangle. In Sinhala, the language of Sri Lanka, there was only one word for all four-sided shapes, and in Swahili, a major language of instruction in East Africa, there was no word for diagonal.

At the same time as many countries were moving to follow the recommendations of UNESCO that early learning, in particular the first three years, should be in the child's first language, the content of mathematics in the primary school was being radically changed due to the influence of projects such as the Entebbe programme, itself heavily influenced by the 'modern mathematics' movement. New topics such as sets and relations, number bases and symmetry were introduced and with them a wider mathematical register (Wilson, 1992). In certain countries, such as Nigeria and Tanzania, bodies were set up under the Ministry of Education to devise new mathematical terms, in Nigeria in Hausa and in Tanzania in Swahili. Lassa (1990) describes how in Nigeria the mathematical register was expanded by either specializing the meaning of words in everyday use, borrowing from neighbouring languages such as Arabic, or assimilation from English through changes in pronounciation ('line' in English became 'layi' in Hausa).

Interesting though this approach may be, it has two major weaknesses. First, it is all very well for an academic body to create words but quite another to get them into use. Second, this approach focuses on the need for expanding the mathematical register rather than on an analysis of the educational and learning needs of the child. There was little investigation of the stage in conceptual development when specific mathematical vocabulary items are helpful or how they should be introduced.

It appears that some second-language learners have a wider mathematical vocabulary than first-language learners. This may be due to the greater frequency with which mathematical descriptors such as 'subtrahend', 'minuend', 'product', 'denominator' and 'numerator' are used by second-language teachers and textbook writers compared with first-language counterparts. Examples drawn from the observation of second-language situations abound; here are just a couple. In a first-year class in a Bhutanese secondary school, a teacher introducing square roots used the word 'radicand'. In the Swaziland Primary Certificate Exam of 1989, children were asked to 'Choose the correct quotient for the following division 10918 ÷ 53'. Why are descriptive words like these used so frequently by second-language teachers? Is it because the lack of linguistic skills in the second language limits the ability to explain and express ideas in several ways? This is clearly sometimes the case. Or is it that the power of a naming word to help concept development is particularly effective for second-language learners? If this is the case then there are implications for all teachers working with second-language learners.

THE STRUCTURE OF LANGUAGE AND MATHEMATICS

Language, however, consists of much more than mere vocabulary. The language used by a mathematics teacher, a textbook writer and an examiner involves the use of particular structures or grammatical patterns associated with the mathematical register. One aspect of language use that has been highlighted over the last decade has been the enormous variablity that can exist in what seem on the surface to be the simplest of situations. Consider the variations of language a teacher may use to describe 12 ÷ 3. Here are six examples, to which you may be able to add others:

> Twelve divided by three
>
> Three into twelve
>
> How many threes in twelve?
>
> Twelve divided between 3
>
> Twelve divided into 3 parts
>
> Twelve apples shared among three children

In the same way as a multiplicity of expressions have been used to explain or 'gloss' mathematical operations, so there is a wide variety and complexity of semantic structures that have been used in the framing of seemingly simple word problems. Riley *et al.* (1983) analysed addition and subtraction word problems and suggested that they could be sorted into three main categories. The first involves a change taking place with the initial data, the second involves a combination of the data, and the third involves comparison of the starting information.

> 1 Abba had three oranges. Then Hamid gave him four more. How many oranges has Abba now?
>
> 2 Abba has three oranges. Hamid has four oranges. How many have they altogether?
>
> 3 Abba has three oranges. Hamid has four oranges. How many more oranges does Hamid have?

They further subdivided these categories by considering the direction of change and the nature of the unknown in the problem. This then gave a total of fourteen distinct categories. Such categorization has stimulated some interesting research into the relative difficulties children have with the different types, into the strategies children use to solve the problems, and into which categories of simple addition and subtraction word problems occur most commonly in pupil's texts (De Corte and Verschaffel, 1991).

Other slight variations in order or tense, in illustration or the presence of superfluous information (Li, 1990), may also affect the performance of children on simple mathematical word problems. These studies reveal the number of variables that can complicate a seemingly simple learning situation. All the evidence supports the view that the level of difficulty of a word problem is related not only to its mathematical content but also to its linguistic form and semantic structure.

It is not surprising that the detailed analysis applied to addition and subtraction problems has not been widely extended to more complex mathematical problems. A good deal of attention has been paid to certain simple grammatical structures that can be readily identified, to problems associated with the words 'more' and 'less' (Jones, 1982; Thorburn and Orton, 1990) and to problems involving logical connectives such as 'if' and 'then' (Zepp, 1982). One other area that has been extensively explored is the complexity of translation of word statements into their algebraic equivalents. The classic example in this field is the problem;

> Write as an equation:
>
> There are 10 times as many students as teachers.

If the natural order of the sentence is followed in translating to the algebraic equivalent, then the algebraic equation that results is $10s = t$. Thinking about the meaning of the sentence or substituting real numbers for s and t reveals that this equation is not the true equivalent of the word sentence. The equation should be $10t = s$.

It is the very structure and grammar of the English sentence that leads to mistranslation to the algebraic equivalent. This and other examples from the growing body of work on the analysis of the semantic structure of mathematical word problems has alerted educationists to the need to be more sensitive to the problems that language can create due to its structure. It must, however, be acknowledged that the research in the past has been more adept at identifying the problem issues than at providing guidance for effective intervention in the classroom.

SEMANTIC STRUCTURE AND SECOND-LANGUAGE LEARNERS

How do the nature, structure and extensiveness of a language affect the way we think in mathematics? Are there some languages that are more suitable for mathematical thinking?

One hypothesis that has been put forward, the Whorf Hypothesis, suggests that the inherent structure of a language influences the ability to think mathematically. One major danger with this view is that it is but a short step to arguing that one language is superior to another and that there are 'primitive' and 'developed' languages. Many students of language would take the view that every language has the potential to evolve in order to meet the needs of its society (Halliday, 1974). Clearly some societies have changed so rapidly in the last hundred years that there exists a vocabulary lag, especially in the technical and scientific registers, but the ability of a language to explain through paraphrase must not be overlooked. In many languages there is a single word for the concept 'the day after tomorrow', a concept for which English has only a set of words which is, however, still perfectly adequate. In the same way, languages without a particular mathematical term can express the same mathematical concept through words or a phrase.

A more constructive approach to the structure of languages is to look at the distinct and different characteristics of each language and how they support or conflict with mathematical reasoning. Consider, for example, the counting numbers in Xitsonga, Fijian, English and Afrikaans:

	Xitsonga	Fijian	English	Afrikaans
1	nwe	dua	one	een
2	mbirhi	rua	two	twee
3	nharhu	tolu	three	drie
6	tseuu	ono	six	ses
10	khume	tini	ten	tien
11	khume nwe	tini ka dua	eleven	elf
16	khume tseuu	tini ka ono	sixteen	sestien

In certain languages, for example Fijian, the structure of the counting words reinforces the place-value of the numbers in a logical and consistent way. In many European languages the counting words, especially in the teens, are inconsistent. This parameter in the language of counting has been investigated in Chinese and English by Fuson and Kwon (in Durkin and Shire, 1991) and they conclude that languages with a regular place-value system like Chinese allow children to add and subtract more easily and more meaningfully. It is interesting to reflect that many teachers have been aware of the difficulties of the English system and have created their own language of 'ten and one', 'ten and two', to try to overcome the problems caused by the irregularity of the English counting system.

The issue that needs to be addressed, however, is not the adequacy or inadequacy of each language but how the choice of language in the classroom may affect the learning of the child. The issue of understanding word problems already discussed has been investigated by Adetula (1990), and his research indicates that children, as one might expect, perform better on word problems when they are presented in the first rather than the second language.

For many mathematics teachers involved in education systems which are dominated by a second language, the big question that still remains unanswered is 'What language should we use to teach mathematics?' This question is often taken out of the hands of educationists and the responsibility for the decision lies with politicians. In many countries the choice of which language to use as the medium of instruction is an historical accident and the decision to maintain it in place is justified on political, logistic and economic grounds rather than as a response to the educational needs of children. There is, however, a subsidiary but important question to which new answers can and should be formulated: when and how should the change be made to learning mathematics in a second language? Within the privacy of the classroom there still remains the potential for a good deal of experimentation and variation with the choice of language to be used and when and how to use first or second language.

THE PLACE OF DISCUSSION IN MATHEMATICS LEARNING

The emphasis on misleading vocabulary and difficult language structure has concentrated thinking on the negative aspects of language. In the last fifteen years there has been a move towards thinking more positively of how children's language can contribute towards their mathematical development. Cockcroft (1982) included 'discussion between teachers and pupils and between pupils themselves' as an essential element of mathematics teaching at all levels, and a more recent HMI report (1989) on primary education comments favourably on mathematical problem solving organized in small groups to promote discussion.

The issue of oral assessment has helped to focus teachers' attention on classroom conversation, and Attainment Target 1 of the National Curriculum of England and Wales also suggests that pupil talk cannot be ignored. Pupils should be 'talking about their own work and asking questions' at Level 1, and both interpreting mathematical information presented in oral form and presenting findings orally at Level 5 (DES, 1991). Unfortunately, although these are the stated targets, much of the assessment of Ma1 is based on children's written recording of their work and the evidence of understanding from children's talk is sometimes overlooked (MacNamara and Roper, 1992).

The importance of discussion is further underpinned by constructivist thinking, namely that children are not passive receivers of a body of knowledge but are actively involved in constructing their own meaning, so the focus of attention should be on how children think and construct their own ideas (see Chapter 3). There is no guarantee that ideas presented by a teacher will be received by pupils in their intended form. Some form of discussion to allow negotiation of meaning seems essential. A pupil struggling to understand a teacher's utterance may resolve the difficulty by trying to pose a question or by disputing a comment made by a fellow pupil.

Vygotsky believed that a child's ability to engage in 'inner speech' develops out of 'egocentric speech'. It is common for little children to talk to themselves, but as the structure and function of this talking develops (between the ages of about 3 and 7) it becomes isolated from external speech and its vocal aspect fades away. 'The decreasing vocalization of egocentric speech denotes a developing abstraction from sound, the child's new faculty to 'think' words instead of pronouncing them' (Vygotsky, 1962). If such a vital role is being suggested for egocentric speech, is there a similarly important role for classroom discussion? Could the cognitive development of our pupils be dependent on their opportunities for negotiating meaning through talking? Vygotsky believed that the cognitive structures of an individual are actually formed through social interaction. Richards (1991) further claims that to act mathematically we need to converse within ourselves, a process which is learnt through participation in conversation with others. In terms of mathematical thinking in the classroom, the processes of conjecturing, testing and amending hypotheses may develop in an individual only after they have been experienced through discussion.

Unfortunately, current thinking on the the crucial role of discussion does not seem to be reflected in the normal mathematics classroom. As Joffe and Foxman (1989) reported:

> An informal survey of mathematics classrooms visited during the course of the development of APU [Assessment of Performance Unit] materials revealed that . . . [it] is more

common to find pupils working alone on individualised schemes or within a fairly formal teacher-directed 'chalk-and-talk' session.

According to advisory teachers consulted before the writing of this chapter, the situation is still much the same. Most teachers are content with well-tried and tested methods and have not the time or energy to contemplate change. Young teachers are more prepared to organize group activities to stimulate discussion, but sometimes their ideas are blocked by others or dampened by pressures to conform. Richards (1991) contrasts 'School Math', with its typical 'Initiation–Reply–Evaluation' discourse, with 'Inquiry Math', which is the kind of discourse used by mathematically literate adults who ask mathematical questions and enjoy solving problems, proposing conjectures or listening to mathematical arguments. Is there need to change mathematics classroom discourse to 'Inquiry Math'?

Language allows thoughts to surface to a level of awareness where individuals can review them and clarify or modify their ideas. Articulating thoughts distances them from ill-defined notions and is hence said to perform a 'distancing' function (Hoyles *et al.*, 1991). Consider the following discussion between P, G and J. They had been sorting three-dimensional objects topologically according to their 'genus' or number of holes. Thus a reel of sellotape had been classified as genus one and a pair of scissors as genus two. The discussion concerns the genus of a multilink cube with a hole in the centre of every face.

P.   It must be six because there are six holes.
G.   No, because one hole is for going in and one is for going out.
P.   So what's the genus?
G.   It must be three. Each hole . . . You exit by the opposite hole.
J.   But I make it twelve.
G.   Why?
J.   Because you can go through the hole from each direction . . . from the outside or from the inside.
G.   But that can't be right because . . . because this sellotape is not genus two. There is only one hole.
P.   Then it must be six.
J.   But when you go in through one hole there are five holes that you can go out from.
P.   Does that make it genus five?
G.   Yes . . . No . . . It could.
J.   If you stretch open the hole at the top and stretch all the sides (faces) so that the whole thing goes sort of flat you get five holes on a sort of plane . . . So it could be five.
G.   Yes . . . OK . . . I see what you mean . . . It *is* five.
P.   Errrrr?

This discussion is a good example of negotiation of meaning. It resulted in agreement between G and J, although P still seemed unconvinced. One very important role of discussion is thus to help to bring to light students' different ideas with the hope that misconceptions will be resolved. How can this be organized in practice?

PRACTICAL CONSIDERATIONS

The aim will be to help learners to take an active part in their own learning and, by putting their thinking into words, to develop their ideas and make them their own. It is worth considering the characteristics of teacher-led and small-group discussion. In teacher-led class discussion the teacher can:

- focus pupils' attention on what is important;
- extend the mathematical language if appropriate;
- provide explanation when needed;
- link ideas with other topics;
- maintain control;
- monitor the contributions of all pupils;

but it will be impossible, especially in large classes, for each member of the class to contribute. The important features of discussion highlighted by Vygotsky and by Richards are pupil contribution and interaction. Will pupils modify their ideas and understanding if they have not had the opportunity to express their thoughts? Group discussion or working in pairs may achieve more pupil contribution.

Many benefits have been suggested for small-group discussion (for example, Bain, 1988; Shuard *et al.*, 1990). Significant claims are that groups:

- allow each learner to take an active part in the learning;
- achieve more than individuals on their own;
- develop oral and language skills;
- promote personal and social skills;
- give participants confidence;
- are motivating and develop positive attitudes;
- minimize teacher interference;
- free the teacher to listen to children to monitor progress, and to assess.

These are very ambitious claims for discussion. Are they actually realized in the classroom?

RESEARCH EVIDENCE ON CLASSROOM DISCUSSION

There have been several studies that conclude that peer interaction enhances the development of logical reasoning through 'a process of active reorganization induced by cognitive conflict' (Perret-Clermont in Cazden, 1988). A good example of such cognitive conflict is revealed in the IOWA team's transcript of discussion between four boys in a first-year secondary classroom. The boys are attempting to describe the journey of an emergency service patrol car that left the 7.4-km post at 12.04 to meet a Renault which had been stuck at the 6.8-km post since 12.02 (adapted from IOWA team, 1977). One boy wanted to draw a line 'vertically' downwards (Figure 7.1a) for the journey of the patrol car. Another suggestion was a line joining the starting position of the patrol car to the position of the Renault (Figure 7.1b). It was the cognitive conflict caused by the boy who pointed out that with these solutions time would have to go backwards that helped to bring new understanding.

Resolving cognitive conflict is not the only benefit of discussion, however. Consider the following interaction (Paley in Cazden, 1988) between two 5-year-olds:

W.   The big rug is the giant's castle. The small one is Jack's house.
E.   Both rugs are the same.

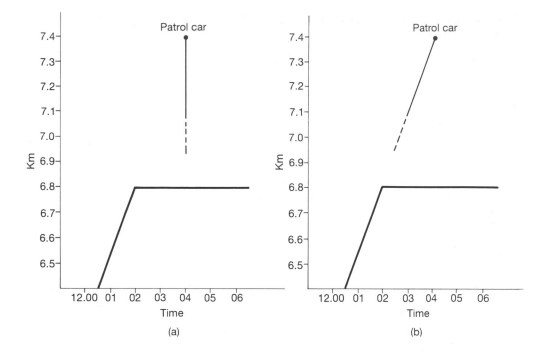

Figure 7.1

W. They can't be the same. Watch me. I'll walk around the rug. Now watch – walk, walk, walk, walk, walk, walk, walk, walk, walk – count all these walks. Okay. Now count the other rug. Walk, walk, walk, walk, walk. See? That one has more walks.
E. No fair. You cheated. You walked faster.
W. I don't have to walk. I can just look.
E. I can look too. But you have to measure it. You need a ruler. About six hundred inches or feet.
W. We have a ruler.
E. Not that one. Not the short kind. You have to use the long kind that gets curled up in a box.
W. Use people. People's bodies. Lying down in a row.
E. That's a great idea. I never even thought of that.

There is disagreement over the size of the rugs but how much reorganization of thinking has taken place is more difficult to establish. Lack of understanding is evident but new ideas are introduced and some seeds of change may have been sewn. What stands out is how each child stimulates the other, and the encouragement offered by E in his acceptance of 'a great idea'.

Mutual guidance and support can be given in many ways, from simple acceptance of a statement to using what has been said to extend the reasoning. Opportunities for learning that arise out of interaction between pairs of pupils in the classroom are analysed by Yackel *et al.* (1991). These include:

- the opportunity to use comments or ideas expressed by another pupil to help develop one's own solution;
- the opportunity to extend one's own conceptualization of a problem in order to

explain why the answer given by one's partner is wrong. For example, Charles was attempting to respond to Connie's suggestion that in 15 minutes' time a clock face showing 3 o'clock would become 6 o'clock, and he based his explanation on Connie's answer. He showed how moving the minute hand to reach 6 o'clock would take far longer than 15 minutes:

> Charles: See, how long does it take to get around a clock? . . . How long would it be? 5, 10, 15, 20, . . . 60. That'd be 4 o'clock. 5, 10, 15, . . ., 60. That'd be 6, no, 5 o'clock. That's 2 hours. 5, 10, 15, . . . 60. That'd be 3 hours. Three hours!

Pirie (1991), however, analyses the script of three girls discussing a problem and draws attention to instances where verbalization constrains thinking:

- one girl's closing statement appeared to prevent another from following up a line of thought;
- a train of thought could have been interrupted and lost through pausing to elaborate for others;
- planning moves and thinking ahead could have been forfeited by the lack of silence.

Telling others may help a speaker to clarify ideas and enable them to be written down subsequently. On the other hand, verbal explanation can be cumbersome and require rewording to bring meaning to others, whereas a pupil on his or her own might quickly jot down an idea in condensed form or using mathematical symbols.

What language do pupils use in mathematical discussion? Relevant discussion will only be possible if participants have some foundation of mutual understanding and shared underlying assumptions (or what Richards (1991) calls an established 'consensual domain'). The more knowledge is shared, the less explicit the language will need to be, for 'shared knowledge allows inexplicit language' (Brown in Pimm, 1987). Do pupils use the language of their teachers or their own everyday language? Pirie and Schwarzenberger (1988) classify the contributions made by pupils in mathematical discussion as 'lacking in appropriate language', 'using ordinary language' and 'using mathematical language'. The confusion between registers mentioned earlier in this chapter also occurs between pupils. Pirie's transcript shows talk moving between different registers and reveals some misunderstanding arising from lack of awareness of the register in use. The everyday use of 'between' in 'the middle one between' referring to a number between 9 and 16 clashed with the mathematical meaning of '7 is between 9 and 16' ($16 - 9 = 7$).

Pupils' understanding of words within a register needs to expand. The word 'number', for example, causes considerable confusion when its meaning embraces fractions and integers as well as natural numbers. This is exemplified by 12-year-old Sarah, who thought $\frac{2}{3}$ was greater than $\frac{3}{4}$ because 'smaller numbers always give larger values'. The pupil was having problems with an expanded understanding of number. The new numbers, fractions and negative numbers, were confused, and because -2 is greater than -3 it seemed that $\frac{2}{3}$ would be greater than $\frac{3}{4}$. Another pupil (Bishop and Goffree, 1986) thought $\frac{1}{4}$ was odd because $\frac{1}{4}$ hour is three five-minute sections. No doubt countless other examples of pupil reasoning and misunderstanding are available to the listening teacher.

THE DYNAMICS OF GROUP DISCUSSION

Groups are known to throw up more hypotheses when problem-solving than individuals (Wall, in Torbe and Shuard, 1982). Solutions are therefore more likely from groups than individuals, though they may take more time. Reasons for the length of time include the following:

- putting your thoughts into words for others must be detailed;
- some pupils tend to think aloud without relating their thoughts to the discussion of the group;
- there are likely to be unhelpful contributions as well as valuable interaction.

Many teachers, particularly in systems which are syllabus-dominated, are worried about the length of time taken up by discussion, but this apparent disadvantage must always be weighed against the possibility of increased understanding. Reteaching the same topic many times is also time-consuming.

Groups seem to throw up ways of reasoning which can be understood by the participants but not constructed by individuals alone. Group discussion thus performs a scaffolding role. Vygotsky (1978) drew attention to the difference between what pupils can achieve when working on their own and when problem solving under adult guidance or with more capable peers, thus emphasizing the effectiveness of the 'scaffolding' role of discussion. Hopefully, group discussion will help more pupils to experience successful problem solving, and if this leads to increased confidence perhaps individual achievement will also improve.

An analysis of studies linking individual achievement and children's interaction in small groups (Webb, 1991) highlights the benefit of giving but not necessarily receiving explanation. The conclusion was that two types of verbal interaction between pupils significantly affect their achievement. Giving elaborate explanations showed positive correlation with achievement, while receiving inadequate feedback (such as the correct answer without any explanation) was negatively correlated with achievement. In general there was little relationship between receiving explanation and achievement. An explanation will only be effective if it links with the level of understanding of the listener. On the other hand, giving explanation always enables the speaker's thoughts to be expressed and reinforced or clarified.

Group work is not likely to be successful if there are poor social relations within the group. 'How pupils think about each other and whether they like each other is important. It's a complex picture and therefore not a very comfortable one for teachers' (Messer, in Munro, 1992). Yet within the group situation it is claimed that pupils learn to listen to each other, to take turns, to value each other and to receive criticism. Whether this happens in practice may depend on

- the successful establishment of classroom norms (Yackel *et al.*, 1991);
- the composition of the group.

There have not been many studies on the effect of personality characteristics on group interaction, but there is some evidence (Webb, 1991) that extrovert pupils give and receive more explanation than introvert pupils, who are more likely to fail to receive responses to their questions. Care must be taken in putting pupils together in groups as there are so many different intentions. Groups might be set up on the basis of friendship, by choice of activity, in order to focus on a particular skill, in order for one of a pair to teach the other, on the basis of equal partnership, in order to develop social and communication skills, and so on. Groups with medium-ability and low-ability pupils, or groups with medium-ability and high-ability pupils, are recómmended as well as homogeneous groups with medium-ability pupils. Groups with a wide range of ability have been found to disadvantage some members and so also have groups confined to high-ability or low-ability pupils (Webb, 1991).

Maintaining the same working groups over a period of time should help to establish group identity, but rotating pupils may be considered an advantage if, for example, there are dominating pupils. Encouraging every pupil to contribute may be difficult but breaking into pairs will produce more participation. Groups may be more successful if each member is given a definite role. 'Information-giver', 'information-seeker', 'note-taker', 'summarizer', 'listener' and 'encourager of participation' are some suggested roles. Certainly group work should provide opportunities for experiencing leadership or developing skills concerned with reporting back or summarizing.

THE ROLE OF THE TEACHER

Allowing pupils to talk involves teachers in a 'sacrifice of overt control' (Mason and Pimm, 1986). It may not be easy to stand back and allow pupils to struggle on their own, or to forfeit the right to evaluate a pupil's contribution and pass on the role of judge to the group. Teachers may feel insecure because:

- they are no longer seen to be in the seat of authority;
- they may have to face questions which they cannot answer.

A major worry for teachers is likely to be how to keep pupils on-task. Given the freedom to talk among themselves, social conversation may be difficult to prevent. The decision of a teacher as to when to ignore non-mathematical talk and when to intervene may depend on the age and ability of the pupils. There have been claims that non-mathematical talk does not necessarily prevent or inhibit mathematical discussion (Pirie and Schwarzenberger, 1988), but other research indicates that off-task contributions have a negative correlation with achievement (Webb, 1991). Lack of concentration is to be expected with an inappropriate or unchallenging task, but there is perhaps need for more research into how long, in general, on-task discussion can be sustained without teacher intervention at different ages and ability levels.

The teacher must decide when discussion will be of value, how to introduce it and how to organize feedback or follow-up. The careful design or choice of tasks, problems or situations which will stimulate mathematical activity and discussion is of paramount importance. Concrete objects are often very successful in forming a focus for the group; for example, hoops to explore the number of regions formed from intersecting circles, counters or cubes to explore the leapfrogs problem, squared paper to investigate nets of an open box from pentomino shapes, and posters like the Möbius band or the Great Stellated Dodecahedron available from the Mathematical Association. Ball (1990) and Brissenden (1988) suggest many discussion activities and games for primary children using buttons, cubes, number cards and dominoes. Mason and Pimm (1986) suggest other ways of introducing discussion. Almost any component of a mathematics lesson (a definition, technique, error or modified result) can be used. Discussion can be 'spend 5 minutes telling your partner why you agree or disagree with this argument' as well as a whole lesson on a planned activity. The variety in use of discussion has been captured on videotape showing lessons being introduced with class discussion before the class breaks into groups – for example, the lesson on modular arithmetic (Open University, 1985) – or lessons incorporating small-group discussion into teacher-guided lessons – for example, Lingard's lesson (BBC, 1988). Small-group discussion may be found to improve subsequent class discussion, and groups can be combined to share ideas or divided into pairs to give more interaction. There is no need to make a final decision for or against discussion or whether to opt for whole-class discussion or groups. Teachers can be flexible and try many different ways of encouraging pupils to express themselves and of allowing inter-pupil talk.

Care is needed over the choice of task. Consider, for example, activities using computers. There would seem to be much potential in mathematical software for generating productive mathematical discussion. Brissenden (1988) examines some software in relation to how well pupils are enabled to build and test ideas involved in the programs. Many good programs introduce problem-solving situations, games and investigations where pupils interact with each other as well as the computer, but little research has been done on which patterns of interaction facilitate learning.

Minor changes in the content of computer programs have been found to affect the relative performance of boys and girls (Light in Munro, 1992). Boys (11–12-year-olds) responded to the problem-solving context of a king searching for his crown while girls succeeded with retrieving honey in a teddy bears' picnic. Other gender differences have been noted by Hoyles (1988). Boys tend to dominate in a computer environment and see less value in communicating with others, while girls appreciate the opportunity to share ideas.

Transcripts of pairs of pupils working with Logo have been analysed in the Logo Mathematics Project. Some individual conceptual development was identified and pupils were motivated to 'feel a sense of ownership of the mathematics' (Hoyles *et al.*, 1991). Through their discussion, pupils generated challenging ideas and discovered more flexible approaches. Comparison of pupils working with Logo and working in a spreadsheet environment suggest that the Logo language gives more support to pupils' language development towards formalization. Generalization can take place by simply substituting a variable for a number. With a spreadsheet, pupils may need to communicate about relationships by pointing at cells. The software environment may therefore affect the style of interaction that pupils are likely to adopt.

In general, an activity which generates fruitful discussion will be motivating and challenging. Finding activities at an appropriate level with wording suitable for the particular pupils may take time. With even the most carefully prepared instructions/questions, some pupils may need help to get started. Further questions may be needed to draw out ideas or vocabulary. The sort of questions Polya (1945) suggests, such as 'Do you know a related problem' or 'Does the problem remind you of anything else?', may encourage pupils to begin to think mathematically. One way of improving discussion in group problem solving may be for the teacher occasionally to act as a role model and solve a problem aloud in front of the class (Richards, 1991). Pupils have to learn to pose the right questions. Perhaps some of these can be copied from the teacher. A typical pupil's explanation might be 'you times that bit by that and then the other by that . . .

and then you see if it's bigger than that'. This is an example of what Brown calls listener-oriented speech (in Pimm, 1987), which is slow in delivery, uses gestures, demonstration and pauses, and often contains many possibly ambiguous pronouns. Pupils need to be guided towards more precise means of expression. They need to be exposed to 'message-oriented language', where information is given with precision, using specific vocabulary and clear syntactic markings like 'because' and 'therefore'. Practice in speaking more clearly and precisely could be given to pupils by asking them to give instructions or descriptions appropriate for a blind person. To improve reasoning in particular, Lochhead (1985) suggests pair problem solving where pupils talk through their solution to a problem with a partner. The role of the teacher in all these situations is to encourage clear, precise dialogue. Some have taken this further and given pupils training in giving explanations to each other (Swing and Peterson in Webb, 1991).

To ensure pupils are learning from activities it will be necessary for a teacher to spend time helping groups to reflect on their activities. The organization of recording of methods and results and the provision for feedback between groups may also need careful planning.

THE TEACHER'S USE OF LANGUAGE

How does a teacher effectively use language to help pupil understanding? Relating school experiences to informal learning at home may be one way. The practical language of shopping, baking, planting seeds, etc., can be brought into the classroom. At the same time as extending the mathematical vocabulary of pupils, the teacher will be trying to meet the pupils at their level, using familiar, natural, everyday language. ('We go to the bank and exchange these 'ones' for a 'ten''; 'Push all the numbers up into the next column'; 'Remember to tidy up your answer'.)

Well-chosen metaphors may sometimes be helpful: 'A matrix is a parcel of numbers', 'Algebra is a shorthand', 'An equation is a balance' are common examples. Providing that pupils are familiar with parcels, shorthand and balances and can focus on the significant points of resemblance, the metaphors may be very enlightening, but Nolder (1991) cites a pupil who was distracted by the unrealistic positioning of cans on the scale pans of a beam balance drawn in a textbook, and this metaphor, while very helpful for solving linear equations, is obviously inadequate for coping with quadratics. Metaphors can also lead to confusion if not carefully applied. The use of 'a' for apples and 'b' for bananas in introducing algebra as a shorthand, for example, completely masks the important concept that a and b represent numbers, and may leave pupils with the idea of letters as objects (Küchemann, 1981). Nevertheless, metaphoric comparison can bring increased understanding. Some primary student teachers in Sierra Leone who were struggling with the idea of square roots were delighted with the suggestion that a root, like a tree root, can be regarded as a line, and finding a square root could be thought of as finding one line of a square of known area. Another example is the enlightenment brought to some pupils by the metaphor of a machine for a function.

If young children are having difficulty in understanding number work, they are often sent back to use concrete materials to link the number operation with action on objects, but according to Walkerdine (1988) what is important is not the link with concrete action but 'the metaphor which allows the task to be located within the framework of a familiar discursive practice'. Her example concerns a child having difficulty with 2 + 3 (Figure 7.2). The teacher does not return to the blocks used initially with the children but changes the metaphor. The circles are called houses, the numbers represent people, and the child has to imagine 'me and my brother and sister leaving one house while you and your sister leave the other and we all go visiting the third house' and 'how many's in that house altogether now?'. To Walkerdine the teacher has a vital role in linking practice and discourse: 'actions on objects do not make sense without a discourse to read them'.

Any account of the role of the teacher in a discussion-oriented classroom would be incomplete without mention of silence. Pupils need thinking time as well as time to express themselves. 'Spend a few minutes thinking before you answer', or 'Think about it and then write down which you consider is the correct answer', may help pupils to make ideas their own. There is also evidence (Carlsen in Richards, 1991) that increasing the 'wait time' when asking questions improves pupils' verbal participation. It is all too easy for whole-class discussion to deteriorate into long questions and comments by the teacher with short answers by pupils. Asking other pupils to comment on an answer or deflecting pupil questions to other pupils can also help to involve more pupils.

There is no easy answer to providing the best learning opportunities for pupils in school, but we cannot afford to ignore the possible benefits opened up by discussion. Activities and situations which lead to fruitful discussion can be introduced alongside teacher-led tasks which may allow more development of language, or role modelling. Nor can we ignore other creative uses of language. The imagination and expression of individual pupils can enrich a lesson when they are given opportunities to create their own mathematical word problems, to explain their own method of solution, or to report on an investigation.

Perhaps the most valuable result of the study of the relationship between language and mathematics has been to sensitize teachers to the language that they and their pupils use in mathematical discourse and to encourage teachers' reflection on how they can help overcome difficulties faced in the task of negotiating meaning.

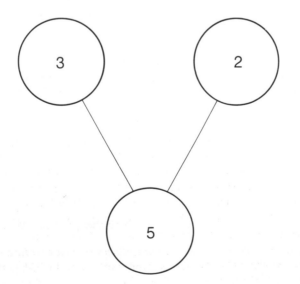

Figure 7.2

Two areas of valuable investigation are suggested for the future:

- more detailed studies of the communicative strategies used by successful teachers and pupils in both first- and second-language systems, for there is much that those in a multicultural society can learn from experienced teachers of second-language learners;
- a deeper study of the role that language plays in the development of children's concepts in mathematics.

Such investigation could help in the search for effective strategies, procedures and activities to increase and refine children's use of mathematical language through all the mediums available to them; through listening, speaking, reading, writing and thinking.

REFERENCES

Adetula, L. O. (1990) Language factor: does it affect children's performance on word problems? *Educational Studies in Mathematics* **21**, 351–65.

African Education Program (1967) *Entebbe Mathematics* Series. Newton, MA: Education Development Center.

Bain, R. (1988) Let's talk maths! *Mathematics in School* **17**(2), 36–9.

Ball, G. (1990) *Talking and Learning*. Oxford: Blackwell.

BBC (1988) *INSET Secondary Mathematics: Five Programmes for Teachers of Mathematics 11–16*. BBC Videotape.

Bishop, A. J. and Goffree, F. (1986) Classroom organisation and dynamics. In B. Christiansen, A.G. Howson and M. Otte (eds), *Perspectives in Mathematics Education*. Dordrecht: Reidel.

Brissenden, T. (1988) *Talking about Mathematics*. Oxford: Blackwell.

Cazden, C. B. (1988) *Classroom Discourse: The Language of Teaching and Learning*. Portsmouth, NH: Heinemann.

Cockcroft, W. H. (1982) *Mathematics Counts*. London: HMSO.

De Corte, E. and Verschaffel, L. (1991) Some factors influencing the solution of addition and subtraction word problems. In K. Durkin and B. Shire (eds), *Language in Mathematical Education: Research and Practice*. Milton Keynes: Open University Press.

DES (1991) *Mathematics in the National Curriculum*. London: HMSO.

Durkin, K. and Shire, B. (eds) (1991) *Language in Mathematical Education: Research and Practice*. Milton Keynes: Open University Press.

Garbe, D. C. (1985) Mathematics vocabulary and the culturally different student. *Arithmetic Teacher* **33**(2), 39–42.

Halliday, M. A. K. (1974) Some aspects of sociolinguistics. In UNESCO, *Interactions between Linguistics and Mathematical Education*. Paris: UNESCO/CEDO/ICMI.

HMI (1989) *Aspects of Primary Education: The Teaching and Learning of Mathematics*. London: DES/HMSO.

Hollands, R. (1980) *A Dictionary of Mathematics*. London: Longman.

Hoyles, C. (1988) *Girls and Computers*. Bedford way papers No. 34. London: Institute of Education.

Hoyles, C., Sutherland, R. and Healy, L. (1991) Children talking in computer environments: new insights into the role of discussion in mathematics learning. In K. Durkin and B. Shire (eds), *Language in Mathematical Education: Research and Practice*. Milton Keynes: Open University Press.

IOWA team (1977) Eleven minutes group work – a transcript. *Educational Studies in Mathematics* **8**, 377–89.

Joffe, L. and Foxman, D. (1989) *APU: Communicating Mathematical Ideas*. London: HMSO.

Jones, P. L. (1982) Learning mathematics in a second language: a problem with more and less. *Educational Studies in Mathematics* **13**, 269–87.

Kouba, V. L. (1989) Common and uncommon ground in mathematics and science terminology. *School Science and Mathematics* **89**, 598–606.

Küchemann, D. (1981) Algebra. In K. M. Hart (ed.), *Children's Understanding of Mathematics: 11–16*. London: John Murray.

Lassa, P. N. (1990) The problems and preparation of mathematical curriculum materials in the mother tongue (Hausa). University of Lagos: *Journal of Educational Research and Development* **1**(1), 66–73.

Li, N. (1990) The effect of superfluous information on children's solution of story arithmetic problems. *Educational Studies in Mathematics* **21**, 509–20.

Lochhead, J. (1985) Teaching analytic reasoning skills through pair problem solving. In J. W. Segal, S. F. Chipman and R. Glaser (eds), *Thinking and Learning Skills 1*. Hillsdale, NJ: Lawrence Erlbaum Associates.

MacNamara, A. and Roper, T. (1992) Attainment Target 1 – is all the evidence there? *Mathematics Teaching* **140**, 26–7.

Mason, J. and Pimm, D. (1986) *Discussion in the Mathematics Classroom*. Milton Keynes: Open University Press.

Messenger, C. (1991) The language issue in lower primary mathematics teaching in Ghana. M.Ed. dissertation, University of Leeds.

Monaghan, J. (1991) Problems with the language of limits. *For the Learning of Mathematics* **11**(3), 20–4.

Munro, N. (1992) Group work complexities. *Times Educational Supplement*, 18 September.

National University of Lesotho (1981) *Report on the Second Workshop on Language in the Mathematics and Science Lesson*. Lesotho: National University of Lesotho.

Nolder, R. (1991) Mixing metaphor in mathematics in the secondary classroom. In K. Durkin and B. Shire (eds), *Language in Mathematical Education: Research and Practice*. Milton Keynes: Open University Press.

Open University (1985) *Secondary Mathematics: Classroom Practice*. Centre for Mathematics Education Videotape, The Open University.

Otterburn, M. K. and Nicholson, A. R. (1976) The language of (CSE) mathematics. *Mathematics in School* **5**(5), 18–20.

Pimm, D. (1987) *Speaking Mathematically*. London: Routledge & Kegan Paul.

Pirie, S. (1991) Peer discussion in the context of mathematical problem solving. In K. Durkin and B. Shire (eds), *Language in Mathematical Education: Research and Practice*. Milton Keynes: Open University Press.

Pirie, S. and Schwarzenberger, R. (1988) Mathematical discussion and mathematical understanding. *Educational Studies in Mathematics* **19**, 459–70.

Polya, G. (1945) *How To Solve It*. Princeton: Princeton University Press.

Richards, J. (1991) Mathematical discussions. In E. von Glasersfeld (ed.), *Radical Constructivism in Mathematics Education*. Dordrecht: Kluwer.

Riley, M. S., Greeno, J. G. and Heller, J. I. (1983) Development of children's problem-solving ability in arithmetic. In H. P. Ginsberg (ed.), *The Development of Mathematical Thinking*. New York: Academic Press.

Shuard, H., Walsh, A., Goodwin, J. and Worcester, V. (1990) *PrIME: Children, Mathematics and Learning*. London: Simon and Schuster.

Skemp, R. R. (1971) *The Psychology of Learning Mathematics*. Harmondsworth: Penguin.

Stockdale, S. R. (1991) A study of frequency of selected cue words in elementary text book word problems. *School Science and Mathematics* **91**, 15–21.

Thorburn, P. and Orton, A. (1990) One more learning difficulty. *Mathematics in School* **19**(3), 18–19.

Threlfall, J. (1993) Private communication.

Torbe, M. and Shuard, H. (1982) Mathematics and language. In R. Harvey, D. Kerslake, H. Shuard and M. Torbe, *Language Teaching and Learning. 6: Mathematics*. London: Ward Lock.

UNESCO (1974) *Interactions Between Linguistics and Mathematical Education*. Paris: UNESCO.

Vygotsky, L. S. (1962) *Thought and Language*. New York: MIT Press/John Wiley.

Vygotsky, L. S. (1978) *Mind in Society*. Cambridge, MA: Harvard University Press.
Walkerdine, V. (1988) *The Mastery of Reason*. London: Routledge.
Webb, N. (1991) Task-related verbal interaction and mathematics learning in small groups. *Journal for Research in Mathematics Education* **22**, 366–89.
Wilson, B. J. (1982) Lims around the world. In *Report of the Second Workshop on Language in the Mathematics and Science Lesson*. Lesotho: National University of Lesotho.
Wilson, B. J. (1992) Mathematics in education. In R. Morris (ed.), *Studies in Mathematics Education 8*. Paris: UNESCO.
Yackel, E., Cobb, P. and Wood, T. (1991) Small-group interactions as a source of learning opportunities in second-grade mathematics. *Journal for Research in Mathematics Education* **22**, 390–408.
Zepp, R. A. (1982) Bilinguals' understanding of logical connectives in English and Sesotho. *Educational Studies in Mathematics* **13**, 205–21.

Chapter 8

Assessing Mathematical Achievement

Dave Carter, Len Frobisher and Tom Roper

THE RELATIONSHIP BETWEEN ASSESSMENT AND THE CURRICULUM

Certain issues were raised in Chapter 4 which relate to assessment; clearly curriculum and assessment are inextricably linked. This is necessarily so since there can be nothing to assess without a curriculum. More importantly, a curriculum must have aims which should be translated into objectives, or pupil behaviours particularly (see Chapter 1). It is the assessment of pupils in relation to the set objectives which informs as to whether the aims of the curriculum have been realized.

Travers and Westbury (1989) see the curriculum at three levels. At the level of an educational system – defined for example by a national curriculum – it is referred to as the *intended curriculum*. This is 'translated into reality' in the classroom as the *implemented curriculum* at the second level. At the third level is the *attained curriculum* – what pupils acquire in terms of a body of mathematical knowledge which includes concepts and skills, the processes involved in doing mathematics, and attitudes toward the subject. Whatever is prescribed in the intended curriculum, teachers and authors of learning resources have to interpret the mathematics in a way that is both meaningful to pupils and in suitable and appropriate contexts. In most systems teachers find that time constraints prevent a thorough coverage, for each pupil, of the intended curriculum; only rarely does it seem possible to explore, with pupils, mathematics beyond that which is prescribed. The study of such mathematics is discouraged as not fitting the curriculum as laid down. Pupils learn mathematics – knowledge, skills, processes – and develop attitudes towards the subject not only through interaction with their mathematics teachers but also by interaction with teachers of other subjects in the wider curriculum, and through their experience of the world outside the confines of institutions. Thus assessing to what extent pupils have attained certain objectives as laid down in the intended curriculum may not reveal all that it is appropriate to know about a pupil's overall mathematical achievement, or from whence such achievement originates.

This immediately raises the question: is there a difference between attainment and achievement? Answers are difficult to find; dictionary definitions are not helpful because it would appear that these words are synonymous. Some authors use these terms interchangeably to mean the same thing. Most authors choose one of the terms and use it consistently to mean, for example, 'how much a child has acquired as the result of specific teaching' (Satterly, 1989). Roby (1990) has the following to say:

> *Profiling and graded certification* both record a pupil's achievement, but the supporting evidence of attainment is frequently omitted from profiling. Before a GCSE grade may be awarded it must be clearly shown how a pupil demonstrated a particular level of attainment.

He is clearly suggesting that, in his view, attainment refers to successful performance in a well-defined task whereas achievement is a summary of all that a pupil has succeeded in demonstrating over a period of time.

Within the National Curriculum of England and Wales (NC) there is an understanding that achievement refers to success in a single task whereas attainment is used to describe success in a group of related tasks. Even the authors of this chapter are not in complete agreement as to the precise distinction between the terms. However, for the purposes of this chapter, we have agreed to adopt the view of Roby (see above).

Evaluation of the mathematics curriculum in order to examine and judge its quality and significance can be carried out in a number of ways. One of the most accessible sources of data is in the form of pupil achievements. However, what pupils achieve may be more directly a result of how successfully the curriculum has been implemented. Indeed, poor pupil achievement is most often attributed to poor implementation; the question of whether the intended curriculum is appropriate was, until the deliberations of the Cockcroft Committee (Cockcroft, 1982), never addressed seriously. This state of affairs can be seen in the present mathematics curricula of many third world countries where the secondary-school syllabuses are still very similar, if not identical, to GCE Ordinary level syllabuses that were once offered in the United Kingdom and exported to these countries. In England and Wales a national curriculum has now been established and in the United States curriculum revision is under way. However, legislation which secures an intended curriculum of quality and significance does not necessarily result in an improvement in teaching quality or pupil achievement.

THE ROLES AND PURPOSES OF ASSESSMENT

Murphy and Torrance (1988) refer to MacIntosh and Hale (1976) when they suggest that scant attention has been paid to diagnosis, evaluation and guidance, although they are regarded by many educationists as more important than selection, grading and prediction, which have been the dominant purposes in the past. Pirie (1989), referred to in Costello (1991), classifies purpose in terms of benefit to pupils, benefit to teachers, and 'satisfying the perceived needs of others'. Denvir (1988), also referred to in Costello (1991), sees the purposes of assessment as being for teaching, selection, evaluation and curriculum control. This particular classification is of interest because Denvir's article was written at the time when the UK government was in the final stages of preparing the NC. With hindsight it is clear that one of the unstated purposes of the national testing of pupils at ages 7, 11, 14 and 16 is to ascertain whether or not teachers are delivering the National Curriculum. The assumption that such testing will actually reveal the extent to which teachers are delivering the National Curriculum should be questioned.

Desforges (1989) offers a simple but comprehensive overview of the purposes of assessment by stating that 'It is generally held that one of the main purposes of assessment is to provide information to help people make decisions.' He elaborates on this by stating that the decision-makers are pupils, teachers, parents, local and national education officers, and 'people who work with pupils when they leave school'. Desforges further suggests that pupils and teachers need diagnostic information, parents need evidence that schools provide sound bases for learning, local and national education officers need information on which to make 'monitoring judgements', and selection procedures for employment or courses in further and higher education are based on information about pupils' achievements. Clearly there is a variety of information required; such variety will certainly not be the result of a single assessment procedure administered on a single occasion. Of necessity, diagnostic assessment is formative and detailed, in order to reveal a pupil's progress, strengths, weaknesses and special abilities. Parents are unlikely to be able to digest such detail but nevertheless need summative evidence that their children are making progress at particular points in time. Because of time constraints, monitoring, whether national, local or within school, is based on aggregated summative evidence. Selection, for the same reasons, is also based on aggregated summative evidence, but such evidence must enable comparisons to be made between pupils. The aggregation of evidence from assessments has its own problems; these will be discussed later in this chapter.

What has not been discussed so far is the motivational purpose of assessment. It is a sad reflection on the content of school mathematics syllabuses, perhaps even on the quality of mathematics teaching, that relatively few pupils are intrinsically motivated by the subject. In order to persuade pupils to learn facts and practise mathematical skills and routines, teachers often resort to extrinsic motivation by way of the threat of tests and examinations. It is often claimed that formalizing assessment in this way provokes pupils to perform at their best. Such a claim may be true for certain individuals but teachers know all too well that for many pupils the stress associated with a formal test or examination is an inhibiting factor. Thus the claim is certainly not one that is worthy of generalization.

On this matter the Cockcroft Report is very clear in the advice it offers teachers (Cockcroft, 1982):

> The form of assessment which is most immediately apparent to a pupil is the marking of written work; this may be routine class work or a more formal test. Such marking needs to be both diagnostic and supportive.

We would go further in suggesting that all assessment, whether formal or informal, formative or summative, should be supportive. This is to say that one of the objectives of any assessment should be to enable each pupil to demonstrate positive attainment. This means that the tasks set may have to be differentiated according to the current level of difficulty at which the pupil is working, or, in the case where all pupils are required to work at the same task, the outcomes in terms of pupils' responses are capable of being differentiated by level of attainment. What is also clear is that, in order for pupils to demonstrate positive attainment, they have to know precisely what is expected of them. It is a widely held view that success is largely a matter of confidence. The Cockcroft Report (Cockcroft, 1982) shares this view:

We believe that there are two fundamental principles which should govern any examination in mathematics. The first is that the examination papers and other methods of assessment which are used should be such that they enable candidates to demonstrate what they do know rather than what they do not know. The second is that the examinations should not undermine the confidence of those who attempt them.

Note that reference is made to 'other methods of assessment' and therefore, by implication, all methods of assessment. Assessment for motivational purposes is a matter of recognizing positive attainment.

NORM-REFERENCED OR CRITERION-REFERENCED ASSESSMENT?

Ridgway (1988) describes the difference between norm-referenced and criterion-referenced tests by stating that the former 'report where a pupil stands in comparison with other pupils who have taken the same test' and the latter 'set out to judge whether or not a pupil has been able to perform some well-defined task to an acceptable standard'. He continues by making the point that the tasks included in a criterion-referenced test must be defined in such a way that pupils and teachers can be in no doubt as to what has been achieved (*sic*). DES (1987) provides a clearer definition of norm-referencing as 'an assessment system in which pupils are placed in rank order and pre-determined proportions are placed in the various grades'. Wiliam (1992) is less precise in his definition of norm-referenced assessment, but makes the point that for a norm-referenced test its ability to stretch out the population over the whole range of scores is vitally important: 'norm-referenced tests need to discriminate'. The car driving test is very often quoted as an example of a criterion-referenced test. However, the criteria of performance for this test are perhaps not as well defined as the learner driver may believe; there is room for subjective judgement by the examiner as to how well the learner has met the criteria – it may not be simply a matter of meeting the criterion or not (Wiliam, 1992).

Norm-referencing a test is likely to be fraught with difficulty. Certain human characteristics, for example the heights of adult British females, or the weights of adult British males, are distributed in such a way that the data is well modelled by the normal curve shown in Figure 8.1; that is, the data is distributed normally. This can be established by taking appropriate measurements for a large, unbiased sample from a particular population. Buckle (1990) makes a very important point in suggesting that it has been assumed that certain other human characteristics which *cannot* be directly observed or measured, for example intelligence and ability, are also distributed normally. This assumption about the distribution of intelligence and ability may be flawed. Buckle does admit that intelligence and ability *may* each be normally distributed, but the assumption that they *are* so has resulted in a widely held view that attainment is also distributed normally. Attainment can be observed and measured directly and therefore there are no grounds for making possibly false assumptions about the distribution of attainment. The discussion so far suggests that tests of attainment should properly be criterion-referenced.

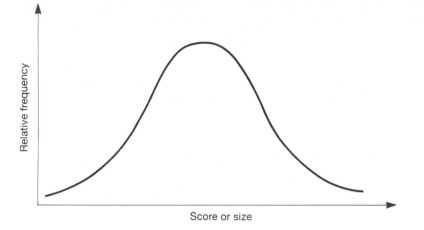

Figure 8.1

PROBLEMS IN MAKING CRITERION-REFERENCED ASSESSMENTS

The content of mathematics in the National Curriculum for England and Wales is laid down in four areas of study which can simply be referred to as 'number', 'algebra', 'geometry' and 'probability and statistics'. There is a fifth area of study which is concerned with the 'use and application', often referred to as the process aspects of mathematics, of the content laid down in the other four areas. The issues arising from the assessment of 'process' are discussed in a later section in this chapter.

Within each of these five areas of study the syllabus is defined within a hypothesized hierarchy of ten levels. Progression through the levels is seen to be attainment-related rather than age-related. This reflects an image of mathematics which is very persistent and one to which many of us subscribe, albeit subconsciously. It is a view which is being challenged by current thinking, particularly through the constructivist perspective (see Chapter 3). It is also challenged via popular writing such as that by Davis and Hersh (1981). In this alternative paradigm, mathematics is viewed rather as an interconnected network than as a strictly ordered set of ideas. Denvir *et al.* (1987) claim that there does exist a hierarchy of difficulty and, by setting criterion-related statements based on this hierarchy, two professionals should be able to agree whether or not a pupil had attained the criterion laid down in the statement. The results from this feasibility study have not gone unchallenged (see, for example, Noss *et al.*, 1989). As professionals we should be concerned about these issues.

However, as far as mathematics in the NC is concerned, for each level within each area of study there are criterion-related Statements of Attainment (SoA, all of which are prefixed by 'Pupils should be able to:'), which in certain cases relate to all the elements of the particular section of the Programme of Study (PoS) but in others only to some of the elements. Pupils who attain some or all of these SoA are attributed that level within that area of study. Within the area of study referred to as 'number', one of the SoA is 'Read, write and order numbers up to 1000'. Three *behaviours*, 'Read, write and order', are defined together with *content*, 'numbers up to 1000'. This statement does not require use of a mathematical *operation*, and no *context* is involved. So it would appear that the objective is well defined and therefore teachers should be able to determine to a high degree of consistency whether or not a pupil has attained the objective. Frobisher and Nelson (1993) discuss the issue of consistency by pointing out that

> Consistency of assessment is not possible unless there exists agreement between teachers about the interpretation which will be applied to Statements of Attainment . . . All valid and dependable criterion-referenced assessment relies upon well defined objectives and shared interpretations.

By way of contrast, consider another SoA in the same content area at the same level: 'Demonstrate that they know and can use multiplication tables'. The content of the statement seems to be ill defined, but reference to the relevant section of the syllabus reveals that 'Pupils should engage in activities which involve: learning and using multiplication facts up to 5 × 5 and all those in the 2, 5 and 10 multiplication tables'. Thus the mathematical content within the SoA seems to be well defined – providing reference is made to the appropriate section of the syllabus. Pupils are expected to demonstrate the behaviours 'know' and 'use'. The problem here is how does a teacher assess whether or not a pupil has reached a state of 'knowing'? What is it to 'know'? Frobisher and Nelson (1993) reveal that the School Examinations and Assessment Council (SEAC) quickly recognized the need to provide a working definition of 'know' as 'provide correct answers using recall (without any obvious counting or computation)'. However, consistent and reliable assessment of pupils' 'knowing' is difficult to achieve. Thus recent NC assessment has relaxed this rigid interpretation. Now pupils are allowed to show that they 'know' if they can respond to a task in 5 seconds. Pupils who can calculate answers very quickly are now assumed to 'know'.

One of the present concerns of teachers of mathematics in England and Wales is that many of the SoA are ill defined in terms of mathematical content and/or behaviour required. For example, 'explore' is a very open operational requirement (see Frobisher and Nelson, 1993). What do pupils need to demonstrate in order that they meet the operational requirement to 'explore'? Until such criteria are presented in a more precise form, consistency of assessment will not be achieved. This raises the question of mastery in relation to consistency of assessment; this is discussed in the next section.

The examples of SoA quoted above are all context-free. Noss *et al.* (1989) argue that it is not possible to have decontextualized 'can-do' statements that have meaning:

> The fact that the 'same' mathematics can be harder or easier depending on the context and manner of presentation is by no means a new discovery, and not in itself crucial to the validity of a given test, as long as relative difficulty between items is recognised. However this issue becomes extremely important if one is attempting to build criterion-referenced topic hierarchies. Notions of decontextualised 'can-do' statements must be strictly meaningless in any criterion-referenced sense. All statements about hierarchies must be conditional on stated contexts and the available tools – they cannot be universal.

ASSESSMENT OF MASTERY LEARNING

Over the past twenty or so years, pupils, and by implication their teachers, have been criticized for 'not having mastered the basics' – whatever that means. Society expects pupils to have mastered certain skills in mathematics such as the ability to recall factual knowledge, to demonstrate that they can use certain skills and routines and to apply these in context. Ridgway (1988) defines a mastery test as a test designed to establish that a pupil has complete command over a particular set of tasks, and in which levels of mastery are set by measures of minimum competency.

Consider again the SoA 'Demonstrate that they know and can use multiplication tables'. Most teachers would accept that pupils who are asked for just one of the multiplication facts defined by that section of the syllabus and supply the correct answer have not met the criterion. Indeed, we would insist that pupils demonstrate knowledge of all possible multiplication facts so defined. Bentley and Malvern (1983) assert that 'nothing short of thorough mastery of the number bonds for the first ten or so numbers is normally considered acceptable'. They further suggest that:

> Other required levels of mastery may not be so stringent. In setting a standard for the adding of fractions, for example, the staff of a school may ask for a success rate of 80 per cent on items which test this skill. This makes some allowances for imperfections in the tests and the pupils. The assessments may occasionally be over-long and later items measure stamina as much as the ability to add fractions.

Satterly (1989) points out that there is no empirical evidence which suggests the degree of mastery required for a particular objective in order that an objective at a higher level be attained. He further points out that there are many routes to the attainment of the same objective and therefore it is difficult to be prescriptive about levels of mastery. Definitions of mastery are likely to be arbitrary.

Frobisher and Nelson (1993) refer to information collected by the Evaluation of National Curriculum Assessment at Key Stage 1 (ENCA1) project in discussing the results of a small survey of teachers' use of mastery levels when making teacher assessments. Although many teachers suggested a mastery level in the range 70–80 per cent, the responses overall ranged from 60 per cent to 100 per cent. There is a real issue here, for as Frobisher and Nelson rightly point out, 'Unfortunately, if consistency of assessment and dependability of outcomes is an objective of national assessment then this variability is unacceptable.' Clearly the claim that a pupil has mastered a particular skill is meaningless unless the level of mastery required is also revealed. The real point at issue is not solely 'how many correct on one occasion', but 'on how many testing occasions'.

The attainment of mastery on a particular occasion does not guarantee that level of performance will be maintained. Sportsmen and sportswomen develop skills to an extremely high level of mastery and through practice maintain that high level. For them practice involves continual self-testing of skills at set levels of mastery. However, mathematics teachers must act with caution and not be persuaded to regard 'mastery of the basics' as being the main objective of the attained curriculum – despite what many critics of educational standards would have the public believe. Satterly (1989) warns teachers of the following dangers:

> If time spent in mastery of a criterion is of paramount importance then one could easily envisage some children taking years to reach a level of mastery attained by others within weeks! Quite probably this is already the case, but rigid adherence to the approach could lock slow learners into an inexorable and morale-sapping ritual, and bore the most able.

ASSESSMENT FOR DIAGNOSIS

HMI (1985), a short but influential book, lays out ten principles associated with assessment of pupils' mathematics. Principle 6 on 'Diagnostic Assessment' starts with the words 'Teachers need to know what pupils find difficult, and why they find it difficult. Without the latter diagnosis any action to remove the difficulty will probably be ineffective.' The vast majority of teachers would give wholehearted support to this statement, but few would be able to claim that they had the knowledge and ability to operationalize it in the classroom. Thus eight years on there are few schools which have taken on board the implications of this HMI principle, and pupils continue to make the same mistakes over and over again as little or no diagnosis of their errors is ever made.

One of the purposes of NC assessment in England and Wales is its formative contribution to methods of teaching. A part of that formative role is a diagnostic element. It is unfortunate that diagnosis has been subsumed within formative assessment, thus resulting in an emphasis being placed on the usefulness of formative assessment for the teacher and not, as diagnosis focuses, on benefits for the pupil. As has already been stressed in this chapter, scant attention is given by teachers of pupils of all ages and abilities to the positive role that diagnosis of pupils' weaknesses can play in improving a pupil's quality of learning, and thereby eventually raising overall standards in mathematical achievement.

Every teacher recognizes that despite teaching of the highest distinction pupils continue to develop misconceptions, encounter difficulties in their learning and make errors in their mathematics. One way forward towards the possible elimination of such occurrences being repeated by every cohort of pupils is to begin to create a distinctive and crucial role for diagnosis in the teaching of mathematics.

This positive role is discussed in the remainder of this section. The discussion is based upon three important principles which every teacher of mathematics should take on board in developing a diagnostic approach to their classroom work:

1. Pupils do not set out to develop misconceptions and produce errors consciously.
2. Difficulties, and consequently errors, occur partly as a result of the performance of teachers.
3. The inconsistencies which exist in the way mathematics is expressed and recorded contribute in no small way to misconceptions, difficulties and mistakes which pupils make throughout their study of mathematics.

Although it appears to be a harsh accusation to level at teachers, there is substantial evidence to show that they are partly the cause of pupils' lack of success. This is not to suggest that teachers are conscious that their performance is in any way inadequate. The majority of teachers, however, would acknowledge that frequently pupils only produce errors in calculations after they have been taught an algorithm which they are required to practise and reproduce at some later date. It is essential for teachers who wish to develop a diagnostic approach to teaching mathematics to recognize when and how misconceptions and difficulties arise and the contribution their teaching may be making to these occurring.

Over the years it has become accepted in primary schools that the order of teaching pupils to add 'tens and units' to 'tens and units' is firstly by the non-exchange (some may prefer the word 'borrowing') algorithm preceding those additions which require the transfer of tens from the units. This is frequently justified on the grounds that the easier algorithm should be taught before what is considered more difficult. (Whether the measure of difficulty is because of the complexity of the mathematics or because of a cognitive factor is seldom mentioned.) Thus pupils readily learn how to do examples similar to this:

$$\begin{array}{r} 3\ 2 \\ +\ \ 1\ 5 \\ \hline 4\ 7 \end{array}$$

The vast majority of pupils experience success with this type of example, as measured by the number of ticks they receive in their workbooks or exercise books. Sometime later pupils are introduced to what are considered to be 'harder' additions. A new or modified algorithm is taught. A few pupils succeed immediately with this. Others appear initially to succeed, but very soon lapse into using the algorithm which worked for the earlier type of addition. Thus they do the following:

$$\begin{array}{r} 3\ 7 \\ +\ \ 1\ 5 \\ \hline 412 \end{array}$$

When teachers see this happening their reaction is to claim that the pupil does not understand place-value. It is true that the pupil who makes this kind of error has not used place-value ideas in the calculation, but it is unfair to lay the cause of the error at the feet of this notoriously difficult concept. The fault lies with the order in which the two types of addition have been taught. When pupils learn an algorithm and have their mastery of the skill confirmed by the receiving of rewards, that is ticks, it is only to be expected that they will resort to the use of the same algorithm in circumstances where it is difficult to see that its use is inappropriate. The blame for pupils making these errors lies with the performance of teachers, who have too readily accepted without question the traditional order of teaching the two types of addition algorithms.

Examples also abound where teachers have seriously misled pupils by vague and imprecise use of language when teaching mathematics. When the area of rectangles is being taught, it is still possible to hear the rule being stated that 'area is length times breadth'. It is undesirable that at some later stage pupils in trying to find the area of a triangle multiply a length by a breadth. The language that teachers use to assist pupils in their learning of mathematics is frequently a barrier to that learning. Teachers who adopt a diagnostic approach become more conscious of the words and expressions which they use when discussing mathematical ideas with learners. As a result they create fewer opportunities for pupils to follow misunderstood, badly phrased 'rules'.

Although mathematics itself may be considered to be a 'logical' discipline, the same cannot be claimed of the way that mathematical concepts and ideas are expressed, both in words and in symbols. There are many examples of which perceptive teachers are aware that illustrate the difficulties pupils face because of the inconsistency of language and symbolic expressions which are used to communicate mathematics.

In reception classes, and at home, children are taught to count objects by touching one object after another and matching the touch with a number word. Thus the following occurs.

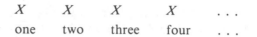

$$X \qquad X \qquad X \qquad X \qquad \dots$$

one   two   three   four   . . .

When children point to the first object this is named 'one'. The second object, although only one object like the first, is named 'two'. This contradiction continues as each successive object, while still being only one in quantity, is given a different cardinal name. The confusion arises because ordinality, that is the order of touching the objects, is used to determine the cardinality of the number of objects so far touched. When children encounter problems with the counting of sets of objects or pictures of objects, diagnosis would seek to establish the cause, of which the above may be only one possibility. Unless the reasons for the resulting counting errors are quickly discovered, little mathematical progress can be made by a child.

At the early stages of learning algebra, many pupils make an habitual error which teachers will readily recognize:

$$a + 2a = 2aa$$

To identify the mistake is all too easy; to hypothesize the explanation for it is more difficult. When such an error occurs, to mark it with a cross and not to talk over with the pupil the reasoning which has led to the error being made is a dereliction of a teacher's role. A possible cause of pupils working in this way is the omission of the '1' as a coefficient of the first 'a'. This is taken as understood by those experienced in the manipulation of algebraic symbols. But to a newcomer to the algebraic scene there is no rationality in its exclusion. A pupil faced with the dilemma of what to do resorts to that which appears to be an eminently sensible method, often based upon previously successful methods. Only in a diagnostic focused discussion with a pupil is it possible with any degree of certainty to arrive at possible explanations for pupils manipulating symbols in this way.

There is a marked tendency to assume that diagnosis of misconceptions and errors is only necessary and appropriate for those pupils who are 'not very good at mathematics'. This attitude results in teachers of 'more able' pupils not adopting a diagnostic approach. Unfortunately, it is this kind of attitude which has led to more and more pupils not succeeding in the subject, and eventually not continuing to study it at a higher level. Every pupil who has had some success with mathematics and as a result continues to study the subject to an advanced level is not immune from such experiences. Teachers of such pupils will recognize this common error, which is to be found in worked exercises and examination scripts:

$$\text{If } y \quad = 3x^2(x^2 + x)$$

$$\text{then } \frac{dy}{dx} = 6x(2x + 1)$$

In an ideal situation any pupil encountering difficulties would have the support of a teacher skilled in the diagnosis of mathematical misconceptions and errors. In turn such a teacher would have available a battery of diagnostic instruments to call upon which would provide assistance in establishing the origins of a pupil's 'rational' misunderstandings and consequent actions. No such teachers are to be found in schools, nor are ready-made, in-depth, intensive diagnostic tests available. This serious deficiency should not be an obstacle to individual teachers adopting a more definite and constructive approach to diagnosis in their classrooms. Pupils' responses should be continually analysed for errors which may throw some light on a pupil's thinking. This should be followed immediately by a discussion with the pupil to root out a likely cause. Until this is done, any remediation is unlikely to succeed. To give a pupil more of the same after a further bout of instruction and explanation is self-defeating; remediation should never *precede* diagnosis.

There is much that a coordinator or head of department of mathematics can do to set the tone for a diagnostic approach within their school. Personal example is of course essential, but not sufficient. Every department should adopt a diagnostic policy towards its teaching of mathematics. Frequent reviews of errors that pupils are making need to be brought to the attention of every member of the department followed by discussion on possible causes. Case studies of individual pupils who are not meeting expectations should take place. These would involve an in-depth look at the work of a pupil since the downward path began. An increasing awareness of situations which give rise to possible misconceptions and errors will result in avoidance of such experiences for pupils.

Extensive assessments such as formal tests and examinations provide pupils' responses which should not only be marked in the conventional way, but analysed for any consistent errors made by an individual or by groups of pupils. Diagnosis of these errors should not only probe pupils' thinking, but also consider teaching methods which may have contributed to the errors. In this way the future quality of pupils' learning will be enhanced and standards will be raised.

It is incumbent on all who wish to improve their teaching of mathematics to embrace a diagnostic approach to their work in the classroom with all learners of the subject. Teachers should consciously aim to make today's failures become tomorrow's successes. The reward will be a recognizable increase in the number of pupils who continue to study mathematics successfully beyond the compulsory period.

THE ASSESSMENT OF PROCESS VIA COURSE WORK

During the early 1960s, a second tier school-leaving examination, the CSE, was set up in England and Wales to provide a more appropriate examination for those pupils not well served by the existing Ordinary level examination. Among many innovations that this examination brought with it was the idea of assessing the work of pupils which was completed outside the examination room; work done during the course – that is, course work. This had as much to do with the idea of continuous assessment and the motivation of potentially difficult pupils as it had with being mathematically innovative. This initiative was later given added weight in the Cockcroft Report (Cockcroft, 1982), which made a very clear statement about the shortcomings of time-limited written examinations:

Examinations in mathematics which consist only of timed written papers cannot, by their nature, assess ability to undertake practical and investigational work or ability to carry out work of an extended nature. They cannot assess skills of mental computation or ability to discuss mathematics nor, other than in very limited ways, qualities of perseverance and inventiveness. Work and qualities of this kind can only be assessed in the classroom and such assessment needs to be made over an extended period.

Previously in the Report, the Cockcroft Committee had provided a very strong hint as to what course work might entail. The now internationally famous paragraph 243 is quoted in full on p. 59. Proponents of this recommendation saw that opportunities for practical work, problem solving and investigational work ought to result in pupil output which was not appropriately assessed by time-limited written tests or examinations.

The examination boards, which control the national testing of pupils aged 16 in the form of the GCSE examination in England and Wales, have tried to encourage teachers to integrate course work into the secondary mathematics curriculum and not to treat it as something to be bolted on at various times during a course. However, Wolf (1990) suggests that the assessment criteria laid down for these national examinations limits the choice of particular tasks or types of task to those which most closely fit the criteria. Thus she suggests that investigations within mathematics offer greater opportunities for pupils to attain the stated criteria than practical tasks or those relating to the real or pseudo-real world. This is based largely on the perception that the latter types of problem are likely to offer one and only one suitable answer and thus are limiting, in terms of the criteria relating to reasoning, making deductions and extensions, and opportunities to develop the problem.

This observation is supported by evidence through In-service Education and Training (INSET), in which one of the authors has been involved. Teachers were asked to set their pupils the same variety of tasks and then to report on their marking of the pupils' responses. There was an overwhelming preponderance of work on one specific task – one which enabled pupils to cover a wider range of criteria than any of the others. While this is clearly efficient behaviour for those teachers working in a bolt-on mode, it indicates that their pupils probably have limited opportunities to attempt practical, pseudo-real- or real-world problems or to work in independent, investigative ways. This is confirmed by the formulaic solutions to be seen in pupils' work. Where class sets of solutions had been returned, virtually every pupil in the class attempted the task in the same way. Seemingly an algorithm for investigating such problems has been applied without consideration of the problem and its own individual features. Indeed, the pupils were applying an algorithm much as if they were doing long multiplication. The justification would appear to be the certainty of pupils attaining some of the assessment criteria irrespective of whether or not they had a grasp of the mathematical import of the task. The 'long multiplication of investigations' has become part of the curriculum in much the same way as content and so falls prey to exactly the same kind of teaching. Burton (1984) summarizes the problem by saying, 'But even where attempts have been made to introduce an emphasis on process into the curriculum, little impact is visible. The reason is partly that once process is enshrined in texts, it becomes content.' She continues, 'An over-conscientious concentration on the content of the mathematics would therefore be expected to obstruct the development of the kind of awareness on which mathematical thinking is based.' It would appear that much the same can be said about an 'over-conscientious concentration' on process.

MacNamara and Roper (1992a, 1992b, 1992c) have also reported on the mathematical activity that goes on unobserved or unrecorded by the teacher or is actually suppressed by the pupils themselves because they do not value it. It could be that, despite the clear necessity for sharing the assessment criteria with pupils, an over-emphasis on criteria on the one hand, and presenting pupils with ready-made algorithms on the other, cause them to suppress the real discoveries that they make because these do not either match the criteria or fit into the algorithm being used. This interpretation is one that requires further research.

Despite these consequences, both real and supposed, which follow from the assessment criteria, there is little doubt that, as far as the GCSE goes, teachers have a shared understanding of the grade that a piece of course work should merit. Grading exercises used as part of INSET activities show that the vast majority of teachers are able to agree upon the grade that should be awarded against set criteria to pupils' attempts at certain coursework tasks. The same exercises have been repeated in Kenya and Botswana. These have shown agreement among 75 per cent of the audience, while in England and Wales the figure for agreement is in the region of 95 per cent for practising teachers and 75 per cent for teachers in initial teacher training.

However, what has to be taken into consideration here is that teachers may be homing in on a 'normative standard' rather than applying the criteria in a rigorous manner. Thus pupils are awarded a particular grade because they exhibit in their work features which a teacher regards as typical of work of a particular standard. This then becomes a reference point for future grading and we have a self-fulfilling system. Wolf (1990) makes similar points, and Roper and Dignan (1988) argue that this is in fact a justified way of proceeding, that what is being arrived at is a professional judgement – one which is informed by the criteria and not arrived at independently. Problems occur when the procedure for making the assessment becomes divorced through custom and practice from the criteria, and post hoc justifications are offered, resulting in corruption of the assessment.

Previously in this chapter we have referred to the five areas of study in mathematics prescribed by the National Curriculum for England and Wales. It is in the area 'use and application' that an attempt has been made by the NCC to categorize the process aspects of mathematical activity by way of SoA. This area is subdivided into three strands: (1) Applications, (2) Mathematical communication, and (3) Reasoning, logic and proof. These strands are seen as being intertwined and yet have separate SoA. The original title of the first strand was 'Making and monitoring choices'. Many teachers would have preferred that this title had been retained when the mathematics curriculum was revised in 1991, since it more accurately describes a strand which focuses upon the ways in which pupils set about and undertake tasks.

The SoA are illustrated by examples, each set in a context, of what pupils might do in order to be judged to have attained the particular statement. These statements are supposed to represent a progression through each strand in a hierarchy of difficulty. There is no empirical evidence quoted by the National Curriculum Council (NCC) to support this hierarchy; but in NCC (1992a, 1992b) the Council reveals its basic conception of progression by stating:

> Progression in using and applying mathematics depends upon progression of each of these following four dimensions.
>
> *Progression in problem-solving and investigational strategies.*
> *Progression in command of mathematical concepts and techniques.*
> *Progression through familiarity with a range of concepts.*
> *Progression in personal qualities.*
>
> Using and applying mathematics provides the experiences through which progression in the four dimensions occurs. The SoA in AT1 are concerned with progression in problem-solving and investigational strategies.

This is an extremely unhelpful definition of progression, showing lack of thought, circularity of definition and an unwillingness to be honest about the problems of defining progression in this area of study. The problems of defining progression in process as opposed to content must be addressed. If process is to be assessed against some hierarchy of statements then there must be a very clear idea of what progression through this hierarchy might mean. Hence the idea of progression is extremely important in the context of process. However, it should be recognized that there is, as yet, no evidence which supports the view that progression in 'process' exists, and if it does that it is the same for every pupil.

It is clear that, from both the GCSE assessment criteria and the SoA within the NC, the content attached to any particular level is a relevant factor in considering the level of attainment in 'use and application'. Personal communication with the NCC suggests that tasks used for assessment in 'use and application' should ideally be set in mathematics content which is within one level of the 'process' level. This is limiting on the range of levels available. For instance, a pupil, having attained Level 5 and no further in each of the content areas, might be capable of offering a proof based upon content designated as Level 5 in the curriculum. However, since proof is seen to be a Level 10 process, the content on which a proof is based should be at Level 9 or 10. This pupil would not, according to the NCC communication, be regarded as having attained Level 10 in 'use and application' due to the disparity between the levels. The processes involved in mathematical activity – comprehend a task, undertake the task, conjecture, make generalizations, offer justifications or proofs for conclusions, communicate findings – have been translated into a hierarchy of SoA. Apparently this progression is to be limited by the level of attainment in content areas. However, Burton (1984) describes mathematical thinking as a dynamic process which is gone through repeatedly. It is thus questionable whether the separate elements of this cyclic process should be used to construct a linear hierarchy. The problem of constraining a dynamic process within a static hierarchy of difficulty is a very real one. Some attempts have been made to overcome this problem – see for example Collis *et al.* (1986).

Analysis of the criteria used to assess course work and the examples used to illustrate SoA within 'use and application' show the same thinking as that used to establish levels and SoA in the content areas. One of the points of criticism offered by Noss *et al.* (1989) is of relevance here. Is such an equivalence justified – not only across areas of content but also across content and process? While acknowledging that there cannot be process without content to process, there can be a great deal of process based upon a small amount of content. Perfectly good and meaningful generalizations and proofs can be given in terms of the structure of the problem without recourse to algebraic symbolism and rules of argument. The efficient mathematical thinkers may be denied access to the levels of attainment which they deserve, sacrificing ability in process to a lack of knowledge or experience of particular content. We should remember that basically what is being assessed is the mathematical thinking of pupils as exemplified by their problem-solving skills; that is, the quality of the processes which they bring to bear upon the problems which they are asked to solve.

The quotation from Noss *et al.* (1989) given in a previous section of this chapter raises the issues of context and manner of presentation of criterion-referenced assessment tasks within the content areas of mathematics. But surely these issues apply equally to process skills and their assessment? Evidence that this is the case comes from Frobisher and Nelson (1993). The research undertaken within the ENCA1 project shows clearly that the performance in aspects of process skills of any particular 7-year-old pupil is dependent on the context of the task. This should not surprise us, but the implication is perhaps a difficult one to accept. This is that statements of what it is expected that pupils can do at particular levels are all relative – relative to the context, the content, the setters of the assessment and their interpretation of the meaning of the statements, and the teacher who must administer and mark or grade the assessment. It is therefore impossible to have an absolute assessment of performance.

Further evidence for the relativity of such statements came from the performance of teachers in an exercise used by one of the authors for INSET. Teachers were presented with separate copies of each SoA from 'use and application' and asked to place them in order of difficulty; that is, to reproduce the hierarchy of levels laid down in the NC document. Naturally this exercise was only possible in the early days of the NC, before teachers had become familiar with it. Nevertheless, even though at that time these teachers had a passing acquaintance with the NC, they were singularly unsuccessful at reproducing the stated, legal order.

RECORDING AND REPORTING MATHEMATICAL ACHIEVEMENT

During the summer of 1992, national tests of attainment for 14-year-old pupils were piloted in some schools in England and Wales. In mathematics these pupils took three one-hour written tests under formal examination conditions. In such a restricted amount of time it is possible to assess only a selection of items in the syllabus, despite the fact that there were four sets of differentiated papers – each set covering four contiguous levels within the hierarchy of ten levels. This means that there was a set of papers covering Levels 1–4, another set covering Levels 3–6, and so on. It should be noted that national testing, for pupils aged 7, 11 and 14, covers only the content areas of the NC – 'number', 'algebra', 'geometry' and 'probability and statistics'. Cutler (1992) reveals that for each set of papers something like 60 per cent of the SoA available within each set of contiguous levels were tested. He further reveals that:

> Those statements of attainment that were included in the tests were assessed at least twice. In order to achieve 'success' with a statement, the pupil had to reach the correct outcome at least once. No one could argue with granting a 'success' if the pupil showed achievement [*sic*] on two occasions, but it is questionable if a single 'success' out of two (or sometimes three) indicates the same expertise.

Given the previous discussion on mastery, one has to question whether a pupil who attains a statement on both of the two occasions tested really has been successful. The statement itself may not be precise enough for a judgement about success to be made.

In the area of study which is referred to as 'geometry', a statement of attainment at Level 6 is given as 'demonstrate that they know and can use the formulae for finding the area and circumference of circles'. Even though SEAC has given a tight definition of the behaviour 'know', there could be a difference in the interpretation of the behaviour 'use'. Presumably 'use' means 'use correctly', but is there a further implication that it also means 'obtain the correct answer'? In the national testing that has been carried at Key Stage (KS) 1, pupils were given credit for 'using correctly' but not necessarily for 'obtaining the correct answer'. We must assume that in national testing the criterion will be applied consistently, but can it be assumed that teachers, when making informal assessments, will share the same interpretation? There may also be some difficulty over the content to be tested in this statement. A strict interpretation of the statement would require that a pupil demonstrates knowledge and use of *both* formulae, but it is possible that only one of these formulae might be tested in national tests. Consider the following two problems:

> *Problem 1*: Calculate the circumference and the area of a circle that has a radius 5 cm. (You may assume $\pi = 3$.)

> *Problem 2*: The area of a circle is 75 cm². Calculate the radius of the circle. Use your answer to calculate the circumference of the circle. (You may assume $\pi = 3$.)

Teachers who are familiar with the content will know from experience that Problem 2 is intrinsically more difficult than Problem 1. Thus what is tested on different occasions may not be at the same level of intrinsic difficulty.

There are three other statements of attainment at Level 6 within the area of 'geometry' – making four in all:

(a) 'Use 2-D representation of 3-D objects';
(b) 'Understand and use bearings to define direction';
(c) 'Transform shapes using a computer, or otherwise';
(d) 'Demonstrate that they know and can use the formulae for finding the area and circumference of circles.'

There is of course an underlying assumption that attainment of each of these statements presents the same level of difficulty. Teachers, in keeping careful records of pupils' attainment, will be able to judge when a pupil has attained all four statements, presumably on several occasions, and therefore can credit that pupil with having attained Level 6 for 'geometry'.

Attainment of a particular level in an area of study may thus take a considerable time. We have previously referred to the area of study 'use and application' in which there are three strands, (1) Applications, (2) Mathematical communication, and (3) Reasoning, logic and proof. These strands embody quite different aspects of mathematical activity. The consequence of this is noted by Frobisher and Nelson (1993) in reporting on the research carried out by the ENCA1 project:

> As the three strands consider different aspects of 'Using and Applying' teachers should not expect pupils to attain at the same level in each of the strands. Indeed the levels could be markedly different as was discovered in the ENCA project, even though only Levels 1 to 4 were assessed. Thus at Key Stage 2 and higher the range of levels of attainment over the three strands could increase as pupils move through the ten levels of the mathematics curriculum.

The national testing of pupils aged 7, 11, 14 and 16 serves to confirm teachers' assessments, or otherwise, for the purposes of official reporting. Whether or not teachers' assessments are confirmed, they are the basis on which teachers make decisions as to a pupil's appropriate level of entry for the national testing. It has already been noted that in the national testing, due to constraints of time and the nature of pencil-and-paper tests, only a selection of statements can be tested. Thus, referring back to the four statements of attainment at Level 6 in 'geometry', if the first and second statements only were tested and a pupil successfully attained these, that pupil would be credited with having attained Level 6 for geometry – irrespective of whether teacher assessment had shown the pupil had attained or failed to attain the third and fourth statements. This point is made very clearly by Cutler (1992) in his critique of the pilot national testing of 14-year-olds in 1992. He goes further by revealing that:

> The level recorded for a pupil within any one mathematics attainment target . . . is the highest level at which the pupil has achieved all the assessed statements. Consequently, as an example, if a pupil is unable to correctly achieve any statements from level 4, but does achieve all the (assessed) statements at level 5, the pupil is recorded as being at level 5 in the relevant attainment target. This means that, unlike the Key Stage 1 system, it is not safe to assume that if a pupil has attained a particular level he or she will also have achieved at all lower levels.

We would wish to add 'unless the teacher, through informal assessments, has recorded evidence to show that the lower levels have in fact been achieved'. Cutler goes on to discuss other anomalies caused by aggregating evidence of pupils' attainment from national testing.

However, this is a relatively minor matter as compared with the issues that arise as a result of the procedures that were adopted in order that a pupil could be given a grade for the subject overall. National testing of the attainments of pupils aged 7, 11, 14 and 16 is meant to provide schools, parents, local and national education officers, indeed the public at large, with reliable evidence about standards in education. For pupils who have sat national tests, the overall grade for mathematics which must be reported to parents is the *average* of the five levels, one in each of 'number', 'algebra', 'geometry' and 'probability and statistics', attained in the national tests, together with the level attained, through teacher assessment, in 'use and application'. Teachers have long recognized the problems resulting from the aggregation of data from school-leaving examinations. One of the problems is that reducing a wealth of evidence to a single numerical or literal grade often persuades an unsuspecting public that this mark is an accurate reflection of a pupil's achievements. In the case of NC assessments, we have tried to show that the attainment of statements may, in many instances, be suspect. The aggregation procedures used serve only to compound these errors and therefore result in disinformation. Clearly the onus is on the class teacher to keep detailed records of pupils' attainments, however onerous that may be. Much information comes by way of observation and discussion as well as marking. The problem seems to be how to record it in a concise form so that it is accessible to other teachers and other schools. A further quotation from Cutler (1992) seems an appropriate way to reinforce this point:

> If parents really want to know how their children are performing in the national curriculum, there is no substitute for the information that flows naturally when parents and schools form a close partnership. Reliance solely on annual reports will not be sufficient.

CONCLUSION

The problems of assessment systems have been discussed in this chapter. All systems have their faults. The difficulty is, in coming to terms with those faults, to arrive at a fair and consistent system which acknowledges the work of the pupil and gives due credit. As has been pointed out, teachers are very capable of developing a normative strategy based upon the criteria or upon past evidence. The dangers of self-fulfilling prophecies are obvious enough in this case, and yet the professional judgement of the teacher in making the assessment against an agreed and consistent interpretation of the SoA or criteria seems to be the only practical way forward. At the heart of this is the phrase 'agreed and consistent interpretation'.

REFERENCES

Bentley, C. and Malvern, D. (1983) *Guides to Assessment in Education: Mathematics*. London: Macmillan Education.

Buckle, C. (1990.) The types, purposes and effects of assessment. In R. Riding and S. Butterfield (eds), *Assessment and Examination in the Secondary School*. London: Routledge.

Burton, L. (1984) Mathematical thinking; the struggle for meaning. *Journal for Research in Mathematics Education* 15(1), 35–49.

Cockcroft, W. H. (1982) *Mathematics Counts*. Report of the Committee of Inquiry into the Teaching of Mathematics in Schools. London: HMSO.

Collis, K. F., Romberg, J. A. and Jurdak, M. E. (1986) A technique for assessing mathematical problem-solving ability. *Journal for Research in Mathematics Education* 17(3), 206–21.

Costello, J. (1991) *Teaching and Learning Mathematics 11–16*. London: Routledge.

Cutler, B. (1992) How is your child doing at school? *British Journal of Curriculum and Assessment* 3(1), 21–3.

Davis, P. J. and Hersh, R. (1981) *The Mathematical Experience*. London: Penguin.

Denvir, B. (1988) What are we assessing in mathematics and what are we assessing for? In D. Pimm (ed.) *Mathematics, Teachers and Children*. London: Hodder & Stoughton.

Denvir, B., Brown, M. and Eve, P. (1987) *Attainment Targets and Assessment in the Primary Phase: Report of the Mathematics Feasibility Study*. London: DES.

DES (1987) *Report of the Task Group on Assessment and Testing*. London: DES.

Desforges, C. (1989) *Testing and Assessment*. London: Cassell.

Frobisher, L. and Nelson, N. (1993) Assessing mathematics. In D. Shorrocks (ed.), *Implementing National Curriculum Assessment in the Primary School*. Sevenoaks: Hodder & Stoughton.

HMI (1985) *Mathematics from 5 to 16*. London: HMSO.

MacIntosh, H. G. and Hale, D. E. (1976) *Assessment and the Secondary School Teacher*. London: Routledge & Kegan Paul.

MacNamara, E. A. and Roper, T. (1992a) Hidden qualities in maths. *Times Educational Supplement*, 24 January, 26.

MacNamara, E. A. and Roper, T. (1992b) Attainment Target 1 – is all the evidence there? *Mathematics Teaching* 140, 26–7.

MacNamara, E. A. and Roper, T. (1992c) Unrecorded, unobserved and suppressed attainment – can our pupils do more than we know? *Mathematics in School* 21(5), 12–13.

Murphy, R. and Torrance, H. (1988) *The Changing Face of Educational Assessment*. Milton Keynes: Open University Press.

NCC (1992a) *Using and Applying Mathematics Book A: Notes for Teachers at Key Stages 1–4*. York: NCC.

NCC (1992b) *Using and Applying Mathematics Book B: INSET handbook for Key Stages 1–4*. York: NCC.

Noss, R., Goldstein, H. and Hoyles, C. (1989) Graded assessment and learning hierarchies in mathematics. *British Educational Research Journal* **15**(2), 109–20.

Pirie, S. (1989) Classroom-based assessment. In P. Ernest (ed.) *Mathematics Teaching: The State of the Art*. London: Falmer Press.

Ridgway, J. (1988) *Assessing Mathematical Attainment*. Windsor: NFER/Nelson.

Roby, B. (1990) Assessment and the curriculum. In R. Riding and S. Butterfield (eds), *Assessment and Examination in the Secondary School*. London: Routledge.

Roper, T. and Dignan, P. (1988) So we decided to do coursework. *Mathematics in School* **17**(3), 18–19.

Satterly, D. (1989) *Assessment in Schools*, 2nd Edition. Oxford: Blackwell.

Travers, K. J. and Westbury, I. (eds) (1989) *The IEA Study of Mathematics I: Analysis of Mathematics Curricula*. Oxford: Pergamon Press.

Wiliam, D. (1992) Some technical issues in assessment: a user's guide. *British Journal of Curriculum and Assessment* **2**(3), 11–20.

Wolf, A. (1990) Testing investigations. In P. Dowling and R. Noss (eds), *Mathematics versus the National Curriculum*. London: Falmer Press.

Chapter 9

Learning Styles and Teaching Mathematics: Towards Open Learning

Geoffrey Wain

INTRODUCTION

In recent years many curriculum subjects have seen the development of teaching schemes which place much more emphasis on the learners to manage their own work. This development has been referred to, among other things, as 'supported self-study', 'resource-based learning' or 'open learning'. For the time being, all of these will be referred to by the the term 'open learning'. Whatever name is given, the intention is that the teacher relinquishes to a large extent those roles concerned with the provision of information, the structuring of the learners' time on task and the sequencing of material, and takes on the general role of manager of learning, the meaning of which will be considered later. The degree to which the teacher retains control of what the learners do each lesson and the extent to which the learners can choose the next task vary from scheme to scheme, but typical of all is that learning is individualized to a very great extent, although often within a group setting. Whole-class teaching is, in general, not used, or used sparingly, and each pupil is engaged in a learning process that is unique to him or her.

The idea of open learning has acquired a particular meaning that implies that the learners have a very high degree of control of their own learning, including making decisions about what to study and when, about how to find a way through the content of the work, and about integrating the topics learned with one another. A more detailed consideration of what is meant by open learning will be given later. The purpose of this chapter is to consider some of the recent developments in the teaching of mathematics, particularly concerned with the individualization of learning, and to explore the extent to which true individualization, or open learning, is possible. There is, of course, a long history of development of pupil-centred learning and teaching. A number of schemes in recent years have been designed to be operated with an individualized approach. Some of these will be considered first before a detailed look at what is meant by open learning is carried out. The idea of open learning will be illustrated particularly with reference to a project carried out in Wakefield in the last few years and in which the author was involved.

KMP AND SMILE

A very early use of a distinctive approach to the individualization of mathematics learning was the Kent Mathematics Project (KMP, 1978), originally given the suggestive title of 'An Auto-instructional Course in Mathematics', a title that was, however, soon abandoned, since 'auto-instructional' gave the impression of intellectual isolation and was not felt to reflect what actually went on in the classroom (Banks, 1971). One of the main motives for establishing the scheme was to provide a means of carrying on the teaching of mathematics in an absence of a sufficient supply of specialist mathematics teachers. Indeed, the problem of coping with the shortage of teachers has been a strong motivating influence in the consideration of such schemes (Gilbert and van Haeften, 1988). The solution was seen to be in providing a scheme that was pupil-focused, teacher-proof, yet externally organized and controlled by specialists. A second project that was concerned with similar issues was originally known as the Secondary Mathematics Individualized Learning Experiment (SMILE), originally based in London schools. Although still known as SMILE, the E is now for Environment rather than Experiment. The structures of the two projects have considerable similarities, which will be described in due course.

KMP and SMILE were certainly among the first in the United Kingdom to attempt to develop individualized learning schemes. Others followed, including the School Mathematics Project (SMP) through its *SMP 7–13* workcard pack (SMP, 1979–80). This while being described as an individualized scheme, at the same time intended that all children should work on the same topic, albeit at different levels, unlike the KMP and SMILE, which allowed children to work side by side on different topics as well as levels. The SMP have refined their general approach in the more recent booklet scheme, *SMP 11–16* (SMP, 1984). Some recent schemes, while retaining a textbook format, include various ways of providing different routes through the material to suit individual needs (Kaner, 1982) or while keeping the class as a whole on the same topic (Blackett *et al.*, 1987).

Some impetus has been provided over the years for the development of individualized schemes by the adoption in many schools of mixed-ability grouping of pupils. Where mixed-ability grouping is used, there immediately arises a need to deal effectively with the problem of differentiation between pupils' needs, speed of learning and depth of understanding. Schemes such as KMP or SMILE are well adapted to coping with these problems. However, as individualized schemes have developed, it has become clear that the form of grouping employed is not really a significant factor. Individualization implies that the relationship between pupils in terms of ability is no longer important and grouping can be carried out using any criteria that have relevance. Mixed-ability grouping requires differentiation of provision; individualized schemes do not require mixed-ability grouping.

KMP and SMILE illustrate many of the aspects of open learning without fully meeting the strict definition of what that implies. The actual definition will not be given at this stage; it will be considered in detail in a later section. For now, the two projects will be described in order to bring out some of the important aspects of their use that bear upon the construction of individualized schemes.

First of all, it is clear that both schemes can be used in a completely individualized way with each pupil learning privately and independently, but, as we have seen earlier, this was not what the originators of KMP intended. Although pupils do spend a considerable amount of time working on their own, they also spend time working in groups of various sizes. It is up to the teacher to determine the proportion of time so spent. There is also no reason why the class should not be brought together for a class session at the teacher's discretion when there is some issue from which the class as a whole can benefit.

The way in which an individual works is subject to the drawing up of what might best be called a contract. In the case of both KMP and SMILE, the contract is in the form of a matrix of tasks selected by the teacher to take the individual's study forward. There are typically up to twelve tasks to be completed in any order, some of which are for individual working, some for pair or group working, and others allowing a free choice of activity for the pupil (Banks, 1976). The tasks usually cover a wide range of content, taking each to a higher level than previously experienced. When the matrix has been worked through, the pupil takes a test question based on each of the tasks completed, thus providing a formative assessment mark, which should be high – that is, in excess of 85 per cent – in order to provide success and motivation. The likely level of the test score is obviously determined by the difficulty level of the chosen tasks, which must therefore be matched carefully to the pupil's ability and attainment to ensure that success is achieved at a high level. It is obvious that such an approach will quickly lead to pupils in the same class working across a wide spread of topics and levels even in a teaching group selected to be of roughly the same standard. Each task has allocated to it a level (from 0 to 10 in the case of SMILE and 1 to 8 in KMP) so that the average level of all the tasks in one matrix provides a general measure of the overall level of performance of the pupil at any time. The process of task selection should ensure that the average level rises steadily throughout the course.

The choice of task is made from a very large bank in each of these schemes. It has already been mentioned that the tasks may be individual or group. They may also be in a variety of forms including workcards, microcomputer software, audio tape material, investigations and short projects. Many of the tasks have been written especially for the projects but use is also made of material produced from a variety of other sources. The full range of material is extensive in format, especially in the case of SMILE, which has always made use of many other schemes and has remained open to alteration and renewal. Thus SMILE has adapted steadily to changing needs and brought in a range of new material throughout its life. In contrast, KMP is in a published form less susceptible to progressive development except at the time of reprint. For any particular task, however, the materials to be used are clearly defined, although there is no reason why pupils should not use other materials if they wish. It is not part of either scheme, it would seem, to seek to encourage pupils to explore a wide range of resources in an open way, although individual teachers would obviously vary in the extent to which they might encourage their pupils to work with materials outside those provided by the schemes. The use of computers is required in some of the tasks although the range of use is rather less than the full potential for computer use in mathematics. In a survey of the use of SMILE materials (Gilbert and van Haeften, 1988) it was reported that the majority of teachers relied very largely on tasks that used only print materials, making little use of other types. The SMILE computer disks do include, however, some imaginative ideas which deserve greater use. Several form a good basis for investigational work and others provide motivating contexts for the practice of routine skills.

As has been mentioned, the teacher is responsible for selecting the tasks for each pupil, and this is done in the case of both SMILE and KMP by referring to a network of activities arranged in topics and in levels of difficulty. A form of the network itself is used as a record of the work done by an individual pupil, by marking on it those tasks that have been completed.

The total number of tasks available through either scheme is far larger than any one pupil could complete, so that the full course of work taken by any individual pupil is highly likely to be different from that of any other pupil. Each pupil, therefore, completes a unique track through the available material, the choices at any stage being dependent upon progress, understanding, interest and application. For the quickest pupils there will be rapid progress through the levels with a large number of the elementary tasks not attempted, while for the slowest there would be a concentration on the easier material, with much of it being used for remedial, consolidation or support purposes. The individualized nature of the work also makes it possible for pupils to pick up their studies after absence without the common problem of having to catch up on missed work.

At each stage the test items completed by the pupil provide a formative assessment upon which the choice of the next matrix of tasks is based. The level of success at each task will determine whether the next task on a similar topic is designed to move the pupil rapidly forward, to provide a steady progression or to enable consolidation or repetition of the work. The pupil's learning is thus subjected to a regular feedback mechanism which is used to match the work set to the pupil's stage of development, aptitude and interests. Obviously the rate at which different pupils work varies a great deal, but a matrix is completed every few weeks and, at that time, an assessment is carried out.

The system as a whole is under the control of the teacher, who makes the selection of the tasks and keeps the record of progress of each individual pupil. The pupil, however, does have much more control of the process of acquiring knowledge and understanding than in a traditional classroom, in which the teacher teaches all the class together and thus closely controls all aspects of the learning process.

Perhaps the most important single feature of the approach embodied in such schemes is that the rate at which a pupil is capable of working is respected at all times. Ideally the rate of progress should be such as to give the pupil both success and/or a challenge in all the tasks set, as well as providing regular feedback to the student from the test results. The pupil should thus be aware of progress and able to appreciate the reasons for the choice of tasks included in each matrix. Ideally there should be an element of negotiation about the tasks to be set between the teacher and the pupil. The pupil also has control over the order in which the tasks entered on a matrix are carried out.

TOWARDS OPEN LEARNING

The description of KMP and SMILE given above is illustrative of a powerful trend in mathematics teaching, which has moved the focus of classroom activity from teaching by the teacher to learning by the pupils (see Chapter 3). Both these schemes and others that are similar are now well established in many schools and there have been some notable successes in their use. Giving pupils more responsibility for their learning does not impair their learning. The question that will now be addressed is whether the pupils can take even more responsibility than is required in these schemes. A natural procedure is to look at the places where the teacher still exerts control and to see whether that control could also move to the pupil, and this will be dealt with below. First, it is helpful to try to define more clearly what is meant by some of the terms that are coming into current use. There have been a number of interesting attempts to give learners more responsibility for their learning and we have seen that these are known variously as open learning, resource-based learning, flexible learning or supported self-study.

In considering the meaning of open learning it is perhaps sensible first to consider in what way learning hitherto may be considered to have been closed. In traditional teacher-led situations there are clearly constraints placed on the learner; constraints that include what is learned, at what time and where. The material to be learned is clearly defined, the means by which it is learned are explicit, the sequence of material is completely in the hands of the teacher, assessment procedures (both during and at the end of the course) are common to all the class, and access to learning is determined by the setting and streaming policy of the school. True open learning might be defined as learning which occurs in a situation where all these constraints are removed. That might be seen as an unrealistic aim, but schemes such as SMILE and KMP have already shown that some of them can be removed. It is also clear that society is moving rapidly to the acceptance of open access to learning as being thoroughly desirable, and open access seems to require a close approximation to open learning if it is to succeed, because it admits of the possibility of wide differences between the individual learners in terms of ability, motivation and purpose.

In a truly open learning situation, respecting the needs of each individual requires that their different styles of learning needs are catered for, and this in turn requires that a range of resources is available to suit different needs. The idea of resource-based learning would seem to have grown out of such considerations. Traditionally teaching has made a heavy use of printed materials, and both KMP and SMILE still do so in practice, even though it would not be expected that they do from analysis of the full range of materials in the two schemes. Openness would seem to suggest that there should be open access to a very full range of resources including all that new technology has to offer.

Of course, one of the most significant resources available to the learner is the human resource of the teacher or, possibly, other teaching assistants. The idea of supported self-study acknowledges this fact by allowing for individual working or working in a group, but with the learners able to draw as necessary upon the support not only of the resources, but of a teacher or other person at appropriate moments.

Approaches using these methods are now common in adult learning, particularly in circumstances where well-defined new skills have to be acquired and where there is, consequently, a high level of motivation to learn. In such cases the learners have a clear idea of what the objectives of learning are and how the newly acquired skills relate to the rest of their knowledge and often also to their employment. Many such schemes are now available but most of them are heavily dependent on written material.

The experience of using schemes such as KMP, SMILE and others is obviously a step towards open learning as defined. The pupils have much more responsibility for their learning, but the teacher still retains control in the essential process of identifying suitable tasks on the network and compiling them into a matrix. This requires the teacher to have a feeling for the network as a whole and for the appropriate balance between topics that are seen as contributing to a coherent mathematics course. The important question is whether school pupils can take on this role for themselves. They will need to be able to find their way round the range of material and also be able to make decisions about which new directions to take. The idea of a curriculum map is a helpful one in understanding the nature of the problem. To a traveller a map is vital, and it may be a convenient analogy to think of the learner as a traveller through the curriculum. Travellers, first of all, need to know where they are on the map and, second, what is the purpose of their journey. Learners also need to know where they have got to and how that position is related to their learning goals. In a great deal of the self-teaching material that is currently on the market, it is typical for a unit of work to begin with the somewhat optimistic statement: 'By the end of this unit you will have learned how to . . .'. Such a statement does, at least, tell the learner what the unit is about so that it is possible to relate the objectives of the unit to the over-all objectives of the learner. Whether the learner does actually succeed in learning what is in the unit is another matter, and depends on several factors such as the degree to which the work of the unit matches the learner's current knowledge and learning style.

The idea of a curriculum map is not new. It is just one of a number of devices that help to make sense of the organization of a curriculum. Such devices include networks of related concepts, hierarchies of learning material, and curriculum flow diagrams. The way that such diagrams are constructed provides valuable information for both the teacher and the learner, but the information provided varies considerably between the different devices. For instance, the network that is used in the SMILE project provides for the teacher a sequencing of material by level of difficulty under a large number of topic headings. This is clearly of value if the only decision to be made is whether a task is more or less difficult than another one. In using this network the teachers need, additionally, to bring to it their own privately possessed network that is concerned with the interrelatedness of the various topics, and they will be able to do this in the light of their own knowledge of the subject and their experience of teaching it. Interrelatedness of topics is not provided by the SMILE network itself. Nor does this network provide any absolutely precise indication of prerequisites required in order to embark upon each task. Interrelatedness and prerequisites could be indicated on a different type of diagram, which would probably be very complex in appearance and use. An example of what a diagram might look like that provides the learner with some idea of the interconnectedness of topics and concepts is provided in Chapter 3 (Figures 3.1, 3.2 and 3.3) and in Orton (1992). To carry the analogy of the traveller and the map a little further, what seems to be needed is identification of the appropriate 'projection' that provides, in a usable form, the information that is required by learners in order to make sense of their journey through the material to be learned.

The properties of the required projection, therefore, would need to include a hierarchical dimension which provides both an indication of the relative difficulty of learning tasks, together with the prerequisites for each, and information about the way in which different aspects of mathematics relate to one another. In addition, however, there is also a need for guidance about the styles of learning that are possible and the desirability of trying to achieve a balance between them. What is envisaged here is guidance for a learner in choosing tasks in such a way as to provide a range of experiences, including working alone, working collaboratively, practising routine skills, applying mathematics to various situations, investigatory work and practical work. The value and importance of each of these also needs to be made clear so that learners can monitor their own progress, experience and achievement.

It is helpful, therefore, to consider a curriculum map as providing three distinct types of information, the first concerning the mathematical topics that might be covered and the interrelations between them, the second dealing with different ways of working and the process skills implicit in them, and the third a general guidance system for making sense of the map itself and providing help in making decisions about where to go next. These three points will be considered again later. Nothing that has been said so far implies anything about the form in which such a curriculum map should exist. The task of the map maker has still to be considered, but before addressing this issue some other points will be dealt with. The aim in devising an open learning system is to give learners control over their own learning. By so doing it is hoped that the learner will not only learn the subject material but also learn how to learn and be more aware of the sources of material and information which will enable future learning to occur.

Learning how to learn requires a knowledge of the resources that can be used. Resource-based learning is one of the names used to describe the teaching developments that are exemplified by schemes such as SMILE and KMP, but the implications of a move to a full open learning situation are that the range of resources made available is far greater than that provided in those projects. This point will be returned to later.

EXAMPLES OF OPEN LEARNING

General practice in primary schools has for a long time exhibited some of the principles of open learning. A recent report commissioned by the United Kingdom government (Alexander *et al.*, 1992) provides a comprehensive statement of good practice at this level. It stresses the purposes of good teaching, and many of these are similar to the ones expressed for open learning.

The main use of open learning in mathematics, however, has been in further and higher education; in particular, in drop-in workshops that have been established in many institutions where the students require uniquely defined programmes to meet their own special needs and which take into account the students' previous experience and qualifications. Use in schools for pupils in the age range 11–16 has been rare in mathematics, although there have been a number of initiatives that have had some success in other subject areas.

A scheme of supported self-study was set up in Holyrood School (Rainbow, 1987) as early as 1982. The scheme was for the whole school and all subjects in the curriculum. It was based on a flexible arrangement of the timetable in the afternoon sessions and was run as an integral part of the work on the government-sponsored Technical and Vocational Education Initiative (TVEI), itself a scheme set up to make use of increased flexibility in the approach to teaching and learning. The use made of this programme by the mathematics department was extremely limited, being confined mainly to remedial work with low-ability children working in a narrow range on content areas for short periods in the day.

There have been other schemes based within the TVEI programme. Tilling (1991) reports the trialling of a so-called 'flexible learning framework' designed to allow schools to analyse their practice in the following three important respects: the management of student/teacher partnerships, of the student use of resources and of student learning pathways. All of these are important in the discussion of open learning in general and all suggest aspects that do need consideration. One of the schools involved used the framework to experiment in mathematics at the A-level, pre-university stage. Allocation of working time was negotiated with a tutor, there were peer-group tutorials, individual or group working and computer-supported work in addition to a few traditional teaching sessions. The findings reported that students gained in confidence in working in groups, using one another as resources for learning and taking responsibility for their own learning (see Chapter 7). Staff reported being more aware of individual needs of the students but, consequently, being concerned at the restraining influence of the current syllabus.

Other case studies can be found in Eraut *et al.* (1991), which provides one of the most detailed accounts of flexible learning and its management, again referring to the use of the flexible learning framework. In presenting the case studies it is claimed that, through them, it is possible to get a glimpse of what the school of the future will look like.

The point made above about the influence of the curriculum itself on the way that learning can take place, and the possible constraining effect of some syllabuses, is an important one. One of the important claims for the use of the flexible learning framework is that it is designed to encourage new ways of learning that are important for the future lives of young people, in which 'knowledge learnt at school will certainly not carry them through, unless they also have developed personal effective skills and the competencies needed to adapt and learn new things' (Department of Employment, 1991). If that is accepted as one of the aims of education, then the methods advocated in open learning take on a new importance and current syllabuses need to be reviewed against these aims.

THE WAKEFIELD OPEN LEARNING PROJECT

The origins of the project on open learning in the Wakefield local education authority (LEA) were concerned with catering for a situation of teacher shortage. As has been mentioned earlier, this shortage motivated the setting up of the KMP. Wakefield schools were partly organized on a three-tier model, with 9–13 middle schools feeding high schools in which the project was planned to take place. It was decided therefore to begin the project with 13-year-olds and to pilot the work in three schools before extending it to other schools and other year groups over a three-year period. Having identified the schools to be involved, there followed a period of materials and staff development.

The materials used were in a variety of forms. Some were designed especially for the project itself, others were taken from commercially produced schemes. The overall aim was to provide for the typical curriculum of the age group but, in addition, to include a range of material that might be described as enrichment or extension material. From the outset an attempt was made to provide for the pupils a curriculum map designed to allow them to find their way around the materials. Initially this was concerned with making sense of the content of the subject. It was decided that this should be done by organizing all the resources under the broad topic headings of number, measure, algebra, graphicacy, geometry and statistics. Thus all resource material was made available in folders that were colour-coded to indicate into which of these categories the material belonged. There were, of course, several items that fitted into more than one area, and this was also reflected in the colour-coding. Thus in making a free choice of activity the pupils were assisted in ensuring that they covered a wide spread of the subject by being made aware of where each activity belonged in a broad categorization.

Each of these main areas of the curriculum was also illustrated in a wall chart designed to provide the pupils with an overview of what was meant by arithmetic, algebra, etc. These wall charts were designed especially for the project and contained pictorial hints of the range of work to be expected under a particular subject heading. Thus the wall chart on number was coloured to match the associated resource material and contained illustrations representing work on fractions and whole numbers, the calculator, and so on. By looking at the chart it was therefore possible to see the range of topics that might be covered and hence provide a guide to selection. Once a pupil or, more frequently, a group of pupils had made a decision about which topic to study, they were able to go to a resource base where a number of packs of material were available from which they could choose something to match their decision.

The packs were all stored in ring-binder files with the appropriate colour-code reference and a title. Each pack provided the pupils with guidance on how to begin the work. The amount of guidance required varied from considerable detail to a brief instruction but, in all cases, was intended to do no more than enable the pupils to get started. Thus at one extreme there was the instruction to use a particular piece of computer software and no other guidance was provided. A particularly popular example of an activity that had such minimal guidance was the Association of Teachers of Mathematics' computer adventure game, L (ATM, 1984). At the other end there was a detailed pack on the theorem of Pythagoras which was designed to guide the pupils through some investigational work leading to practice of quite traditional examples. It is not possible to detail all the materials here, only to give a flavour of what was available.

Once a choice of activity had been made, those involved in it would, typically, have work for a period of anything from a few lessons to several weeks. During this time the pupils were free to use a wide variety of other resource material including books, mathematical dictionaries, apparatus, videos and computers. The place for work was not confined to a single classroom. The usual arrangement was to combine classes in adjacent rooms and to allow the use of corridor space and, for some activities, other areas of the school as well. All the rooms contained a collection of the resource packs, a library of reference books, computer software, computers and a store of equipment. The pupils were expected to make free use of all resources, which, in practice, meant that there was on occasions some initial guidance from one of the teachers available.

The working unit was, usually, a group of pupils, typically three or four in size. In practice the groups worked in a wide variety of ways. For instance, it was possible for a group all to be engaged on the same task or different aspects of the same task in a co-operative way, but it was also quite likely that an individual within a group might work independently at something which he or she felt he or she needed to learn at that time. There were also occasions when the teacher decided that it was appropriate to bring the class together as a whole and organized the groups to work at a single problem, or at different aspects of the same problem, or at different activities arising out of the same item from the syllabus. The choice of which way to work was agreed with the pupils and a class activity of this sort would almost certainly arise out of the current work of some the groups working independently. Teachers found that for the 13-year-old pupils most of the work took place in groups and was determined by group decision. Further up the school the tendency was for more whole-class work to be carried out, even though the groups were still an important feature of the organization.

During work on a particular activity the pupils were required to record what they did. The record took a wide variety of forms, ranging from a traditional-looking exercise book full of worked examples, as in the case of the Pythagoras activity referred to above, to a collection of rather untidy notes in the case of L. Inevitably the pupils had a wide choice of the form in which recording took place – there were few situations where it was a requirement to produce a set of worked examples as in a traditional classroom. Of more importance in the design of the project was the requirement for the pupils to review what they had learned. The form of the review was designed to allow reflection on their own learning and to aid them in the choice of what to study next. This process of review has undergone change as the project has evolved and lessons have been learned about the most effective forms of working.

At the beginning of the project, the pupils were asked to meet in their group after completing an activity to discuss what they had learned and to keep a record in a student review booklet written specifically for the task. After discussion, each pupil was required to fill in a review on the particular activity just completed, highlighting what had been learned, the difficulties that had been encountered, the direction of possible follow-up work, the help received, and their feelings and attitudes about the work. Later in the booklet, in sections on number, measure, etc., pupils were asked to list any new words that they had met and now understood and new facts and skills that they had acquired. It was realized in practice that the pupils found this approach to review quite difficult, and changes were made in the light of experience. In particular it was found difficult to identify a 'new' word, fact or idea. In retrospect it is apparent that this is so because students acquire new ideas over a period of time, which can be very long, so that it was difficult for them to pin-point the moment of revelation. Thus the format of the review booklet was changed and pupils were asked instead to comment on the facts, ideas and words encountered in the work, and also to begin the construction of a concept or curriculum map that provided interrelationships and links between the different aspects of the task carried out. Mention has already been made of the variety of possible 'projections' that might be used to form a concept map, and the pupils' natural 'projection' would seem to be a very significant one. Pupils were therefore asked to produce their map in any one of a number of forms, including representing their understanding in the form of a poster, writing a summary of their work for the purposes of revision, reporting on what they done to the whole class, or writing an advert extolling the virtues of the topic covered.

From time to time a teacher would sit with a group, or groups, if appropriate, or even work with the whole class, to discuss the work that had been done in order to help the pupils make sense of their overall progress. In general the pupils seemed to welcome these periods of reflection, in which they could relate the material they had encountered to that covered by others and with the illustrations on the wall charts. The complexity of their own developing network of understanding of the subject would appear to have been important in assisting them to make decisions about future work. There seemed to be, however, no accepted best approach to giving this particular form of understanding, but the experience of the project has suggested some useful strategies. Discussion in a group allied to recording in progress books appears to be helpful. The construction of a curriculum map of a particular part of the subject was often achieved through the combined efforts of several groups of pupils all contributing from different points of view. The subsequent discussion to reconcile the differences would seem to be valuable in revealing new possibilities for study as new insights into the interrelatedness of various topics emerged. An example of a map of the kind produced in these discussions is given in Chapter 3 (Figure 3.2). It will be seen that there is considerable detail (and some misspellings!) and evidence that the pupils were beginning to see the subject of mathematics in a way that is not normally available at all in a traditional approach to teaching. The illustration given owes something to the wall charts used to motivate an understanding of this kind, but it contains linkages between topics which have been created by the pupils themselves.

Diagrams of this kind were made available to other pupils and provided a new way of guiding choice of future activities. One of the important outcomes of such work was to provide a clearer indication of where choices should be made in order to provide adequate coverage of the whole of the content of the subject. Discussions enabled key issues about topics to be made apparent so that the pupils could develop an understanding of where their learning should be concentrated in the future.

It is helpful, as has been suggested earlier, to conceptualize the work at three distinct levels. First, it is necessary to provide for pupils access to the subject matter of mathematics. A key question here is what limits should be put on the choices available and hence the extent of coverage of mathematical topics. Where there is a well-defined national curriculum there is, of course, the temptation that the contents made available would be exactly those in the curriculum. This would be against the spirit of open learning, where development of knowledge and understanding should, itself, have openness. The National Curriculum of England and Wales (NC) does have the potential for a certain degree of openness through projects in using and applying mathematics, as well as through work on what are known as cross-curricular themes. In an ideal world it might be hoped that a pupil could be allowed unlimited access to the subject provided that the needs of the examination process are met. For very able pupils this would seem to be a wholly sensible approach, especially in cases where a pupil has high talent in the subject. Traditionally the subject has been severely limited by the syllabus, in both its content and the order in which is presented, which for the ablest must always have been a constraining factor.

The work of the project should also be seen in terms of the opportunity provided for pupils to learn in a variety of ways. One of the most often quoted extracts from the Cockcroft Report (Cockcroft, 1982) is paragraph 243 (see p. 59), which proposes that good teaching of mathematics should include a range of approaches including practice of skills, applications of mathematics to other subjects, discussion, problem solving, investigational work and exposition. The Wakefield project, assessed against the proposals in this paragraph, would seem to come out very well. Of course the extent to which it does is determined by the type of learning material available, but the spirit of open learning in itself suggests that a wide range of material must be made available, by definition as it were, and it would be inappropriate not to provide all these approaches. Some indication has been given earlier of the different types of activity that have been supplied. In practice, the range has been extensive and access to it by the pupils has been made relatively easy through the mechanisms of choice that they have been able to use. The project would seem to have avoided the situation that was the source of one criticism of programmes such as KMP and SMILE, namely that, in practice, the pupils used mainly written materials, usually in the form of workcards, while other types of material were seldom used. The Wakefield project had a limited use of workcards in the usual format. Where the choice of what to do is in the hands of the pupils, there may well be a likelihood that the choices made will be from a much wider range of resources than the teachers would naturally use, because of the teachers' background, expectations and general conservatism.

The third aspect to be considered is that of the way that pupils are enabled to direct their studies, and it would seem that the various strategies, described above, that lead to the development of a curriculum map by the pupils have considerable potential. The overall view of the subject that is fostered by such strategies is an important part of the learning process, and potentially can enable pupils to take away with them from their schooling a better understanding of the subject than hitherto, and a basis for understanding where future study of the subject might be appropriate to them. For the purposes of examination it is also important that the curriculum map relates to the curriculum to be assessed, providing the learner with a coherent body of knowledge rather than a fragmented collection of skills. Providing pupils with a 'director system' that allows them to acquire an integrated appreciation of the subject, awareness of what they do and do not know, and a way of choosing what to study would seem to be a very reasonable aim of teaching mathematics that provides for an integrated appreciation of the work.

CONCLUSION

Mention has already been made of the importance of exploiting a wide range of resource material in devising a truly open learning system and of the fact that in both SMILE and KMP teachers tended to use only the printed materials. Evidence such as this would suggest that teachers of mathematics are very traditional in their choice of teaching materials. Evidence from the Wakefield project shows clearly that non-print resources are very attractive to pupils and that the pupils readily make use of a wide range of reference material. The traditional approach of teachers of mathematics – to base all their work on a published scheme, usually in the form of a textbook – dies hard, and adapting to a resource-based approach requires a considerable shift of attitudes.

One particular consequence of a widening of the resource base is that a range of material will be available to the pupils that will take the subject outside its traditional definitions and into areas that may well be cross-curricular. This suggests a new aspect of openness, that is, subject openness or the removal of subject boundaries, which, in the case of mathematics, have been particularly strong.

Bernstein's analysis (Bernstein, 1975) in terms of what he defines as the collection codes and integrated codes is valuable here. In the collection code the acquisition of knowledge is seen as taking place within isolated and insulated curriculum subjects. Teachers have strong subject loyalties, and knowledge within each subject area is organized within a hierarchical structure. Thus knowledge has to be acquired in a clearly defined order laid down by subject-powerful examining committees, which do not have to consider the possibility of relating to each other across subject boundaries. By contrast, the integrated code acknowledges that learning does not, of necessity, respect subject boundaries, but that combinations of hitherto separated and apparently disparate fields of knowledge may give important learning experiences. There is an emphasis on the pupils' determination of what is learned; teachers co-operate across subject boundaries, and control is in the hands of the teachers and learners together. Bernstein makes the point that the collection code 'is capable of working when staffed by mediocre teachers, whereas the integrated code calls for much greater powers of synthesis, analogy and far more ability to both tolerate and enjoy ambiguity at the level of knowledge and social relationships'. True open learning would seem to be closely allied to Bernstein's integrated code rather than to the collection code, and so it is ironic that the motivation for developing open systems, from KMP onwards, has been to counter the problem of the shortage of teachers of the subject. It is also interesting that considerations of the link between mathematics and other subjects do not appear to have emerged from many of the schemes referred to above, in which mathematics was rarely fully integrated into open learning in the way that some other subjects were. The Wakefield project experience suggests that openness cannot be confined to one subject but demands a whole-school policy for its implementation.

One further matter that demands comment is the role of new technology in the mathematics classroom. This issue is considered in considerable detail in other chapters from a number of points of view. As far as the idea of open learning is concerned, there are important consequences of the widespread availability of computers. The use of Logo has been written about extensively, and it is clear that computer environments of this sort provide an openness of the kind that has been discussed. The power of the computer to enable the learner to explore beyond the immediate context of many problems makes it possible for the learners to bring openness to their work in a way hitherto undreamed of. Some of the software produced for the SMILE project has just this quality of openness. It would also be sad to ignore the potential of home-owned computers to provide the learner with activities which contribute to mathematical understanding. This point is also discussed in Chapter 2. What would seem to be important is to research the uses to which pupils put computers both at home and at school when they are free to use them in their own way in any mathematical context.

REFERENCES

Alexander, R., Rose, J. and Woodhead, C. (1992) *Curriculum Organisation and Classroom Practice in Primary Schools*. London: DES.
ATM (1984) *L – A Mathematical Adventure*. Derby: Association of Teachers of Mathematics.
Banks, B. (1971) The 'disaster kit'. *Mathematical Gazette* **LV** (391), 17–22.
Banks, B. (1976) Free choices. *Mathematics Teaching* **75**, 17–20.
Bernstein, B. (1975) On the classification and framing of educational knowledge. In M. F. D. Young (ed.), *Knowledge and Control*. London: Collier Macmillan.
Blackett, N. *et al.* (1987) *NMP Mathematics for Secondary Schools*. Harlow: Longman.
Cockcroft, W. H. (1982) *Mathematics Counts*. London: HMSO.
Department of Employment (1991) *Flexible Learning: A Framework for Education and Training in the Skills Decade*. Sheffield: Department of Employment.
Eraut, M., Nash, C., Fielding, M. and Attard, P. (1991) *Flexible Learning in Schools*. London: Department of Employment.
Gilbert, L. and van Haeften, K. (1988) *Teacher Shortage Subjects: The Contribution of*

Supported Self-study and Other Alternative Approaches to the Teaching of Mathematics, Physics and CDT. London: National Council for Educational Technology.

Hodgson, V. E., Mann, S. J. and Snell, R. (eds) (1987) *Beyond Distance Teaching – Towards Open Learning*. Milton Keynes: Open University Press.

Kaner, P. (1982) *Integrated Mathematics Scheme*. London: Bell and Hyman.

KMP (1978) *Pupil/Teacher Materials: Levels 1–8*. London: Ward Lock Educational.

Orton, A. (1992) *Learning Mathematics*, 2nd edition. London: Cassell.

Rainbow, B. (1987) *Making Supported Self-study Work*. London: Centre for Educational Technology.

SMP (1979–80) *SMP 7–13*. Cambridge: Cambridge University Press.

SMP (1984) *SMP 11–16*. Cambridge: Cambridge University Press.

Sutherland, D. (1990) *Open Learning in Mathematics: Report 1989–1990*. Wakefield: Wakefield Local Education Authority.

Tilling, M. (1991) *The Flexible Learning Framework: A Trialling Report*. York: York Education Centre.

Waterhouse, P. (1988) *Supported Self-Study: An Introduction for Teachers*. London: National Council for Educational Technology.

Chapter 10

Problems, Investigations and an Investigative Approach

Len Frobisher

INTRODUCTION

For many teachers, problems and investigations are thought of as a recent innovation in the mathematics curriculum. However, their history can be traced back for many centuries. The present movement, which has seen a greater emphasis placed upon the role of problems and investigations in the curriculum, was initiated in British schools in the early 1960s by the efforts of a small but enthusiastic group of teachers who belonged to what is now called the Association of Teachers of Mathematics (ATM). At about the same time, some of those involved in the training of teachers in colleges and universities, the Association of Teachers in Colleges and Departments of Education (ATCDE), attempted to move the debate about problem solving into the field of what is now popularly known as 'investigations'. At that time these were referred to as 'open problems'; 'it is the exploration of these more open problems which we feel is the essential characteristic of real mathematical activity' (ATCDE, 1967). This was a radical approach to using problems as part of the learning of mathematics, and the notion of presenting children with such open problems was fully supported by the ATM. They believed that in this way children could be assisted to make mathematics their own, as they would feel partly in charge of their own learning.

At about the same time, changes in the approaches to teaching mathematics began to appear in other parts of the world, although the directions such changes took were not always the same. In the United States a more extreme and systematic development of the role of problems in the curriculum has taken place over the last thirty years, culminating in the National Council of Teachers of Mathematics (NCTM) publication, *An Agenda for Action*. In this visionary statement of the future direction of mathematics in schools there appeared the recommendation that 'problem solving be the focus of school mathematics in the 1980s and beyond' (NCTM, 1980). The view held by many mathematics educators in the United Kingdom did not entirely match this vision, as they saw the solving of problems as only one small part of a much wider arena where the solution of a problem took a minor role compared to that of the processes involved in the investigating of the problem.

Although the elevation of problem solving and investigations to a higher status in the mathematics curriculum dates back to the 1960s, for many years its progress in Britain was limited to secondary schools. It was not until the Mathematical Association (MA) initiated its diploma courses in mathematics education in the 1970s that primary-school teachers were made aware that such changes had relevance for them. However, for most teachers the first reference to problem solving and investigations in the mathematics curriculum came with the publication of *Mathematics Counts*, which has become more popularly known as the Cockcroft Report (Cockcroft, 1982). This Report referred to problems and investigations as essential elements of the teaching of mathematics with all ages and abilities. The aim was to change the role of children from that of passive recipients of transmitted knowledge to that of active participants in their own learning.

Secondary-school mathematics teachers received a forceful nudge towards incorporating problems and investigations into the curriculum when extended course work was introduced as a requirement of the GCSE. The two ideas were not stated in these terms, but were referred to as part of 'two basic categories of open-ended task, called practical and investigational' (Wolf, 1990). This was quickly followed by the short but influential booklet, *Mathematics from 5 to 16* (HMI, 1985), its pages pervaded by the theme of an investigative approach to learning mathematics. It also stressed problem solving and investigations as particular aspects of this approach in points 9 and 10 in the 'principles of classroom approaches'.

In the late 1980s, primary-school teachers received a similar impetus towards incorporating problems and investigations into their teaching by the Primary Initiatives in Mathematics Education (PrIME) project, directed by the late and highly respected Hilary Shuard. Today it is safe to claim that, at the very least, the vast majority of teachers are aware of the significance of problems and investigations in the mathematics curriculum. Unfortunately, after over thirty years, it is not possible to assert with any degree of confidence that positive action by teachers to include them as a natural part of children's mathematics is universally taking place. For waverers and doubters, the National Curriculum of England and Wales (NC) has provided reassurance by including many aspects of problem solving and investigations in the programme of study for Ma1 ('Using and applying mathematics'). The non-statutory guidance for mathematics produced by the National Curriculum Council (NCC, 1989) sought to show how problems and investigations were a cardinal feature of the NC. In a similar way the NCTM's *Curriculum and Evaluation Standards for School Mathematics* (1989) has placed problems and investigations at the centre of the mathematics curriculum in the USA for all children.

There is, however, a serious barrier to the original intentions of the early 'problem-solving and investigation' reformers being implemented. This is the fact that many and varied interpretations are being given to the two concepts. For every interpretation there is a different classroom method of teaching investigations. It is doubtful whether it was ever the intention of the members of the ATM and the ATCDE that investigations or open problems should be taught as if they were topics in the curriculum. That, however, is the way curriculum innovation becomes debased as individual teachers try to find their own way of implementing change.

Uncertainty about the differences and distinctions to be made between 'problem solving' and 'investigations' in mathematics is likely to contribute to misplaced efforts in classrooms throughout the world. Until mathematics educators present an agreed and consistent approach, children may continue to be confused as to teachers' intentions. There is little doubt that a great deal of overlap exists between problems and investigations, but as yet there is neither agreement nor consistent use of the terms in the community of mathematics educators. It is not, as some writers would have us believe, that it is simply a matter of semantics (Costello, 1991). If teachers are unclear about the distinction to be made between problems and investigations, then children can have little confidence in instructions which suggest that they should 'investigate the problem', or 'explore the investigation'.

WHAT IS A PROBLEM?

Which of these are problems?

1. $345 + 534 = ?$
2. Expand $(a + b)(2a - b)$.
3. How many squares are there on a chessboard?
4. Classify polygons according to their angles.
5. Plan an interschool football tournament.
6. There are five people at a party. If each person shakes hands with the other guests, how many handshakes will there be? Generalize.
7. Investigate the addition of odd and even numbers.
8. Explore polygonal numbers.

The reader should attempt to answer each question above before proceeding any further. Then, on concluding reading the chapter, return to the questions and reconsider the responses in the light of what has been read.

For many years, throughout much of this century, there has been a commonly accepted view of what a problem is in mathematics. This view has been based on what is now regarded as just one particular kind of problem which has been part of mathematics curricula from the earliest times, namely 'word problems'. These are problems which attempt to place mathematics in a psuedo-real context and are presented in textbooks through the medium of the written word. They are a very convenient way of presenting mathematics to children in a form completely different from the multitudinous sets of exercises which were also to be found in many older textbooks. Here is a typical word problem involving the calculation of fractions:

> There are 80 books on a shelf.
>
> $\frac{3}{8}$ of them are non-fiction.
>
> How many books are fiction?

Word problems are excellent classroom examples of the task-environment, goal-oriented view of problems. In a word problem, a task or situation is presented in words, and a question is asked which sets out the goal that the solver has to attain. In most instances the algorithmic skill needed to solve the problem will already have been taught and practised. For many years, teachers have taught mathematics problems in the belief that a problem has a goal that a child has to achieve using previously learned knowledge of the procedure needed to reach the goal. All that a child has been required to do besides performing a calculation was to decide which algorithm should be used. This is not a description which teachers consciously express, but it is one which has found favour in many classrooms for over a hundred years. Indeed, in many countries throughout the world such a definition would still be openly provided and justified in relation to the kind of problem which children are taught and by which their mathematical ability is often assessed.

More recently there has been a noticeable shift in emphasis in definitions of 'problem'. Reference is now made to the position and state of the child who may be required to solve the problem. Thus Reys *et al.* (1984) claim that 'A problem involves a situation in which a person wants something and does not know immediately what to do to get it.' In this definition there remains reference to a situation and the end goal, but solvers are now personally involved in that they desire to achieve the goal. Pupils could be asked, for example, to design a box to hold a birthday present. To Reys *et al.*, this would be considered a problem if the pupils really did wish to design a box. Should it not be the aim of every teacher that children wish to achieve the solution of a problem for its own sake rather than because they wish to please their teacher? How frequently are children observed solving problems from textbooks – problems which they have no intrinsic desire to solve? They do so only because that is what one does in a mathematics classroom. All too often such textbook problems only involve children in determining and then performing the appropriate operations in order to achieve the goal. There is very little implicit interest or challenge; they are someone else's problems, and the desire to solve them evaporates as the problems become routine (see Chapter 3). When this occurs the problem is no longer a problem for the child, as it has now lost its inherent worth to the individual.

Increasingly the focus of problems in schools has moved towards the process or method of reaching a solution. The search for an appropriate path which will lead to the goal has assumed much greater importance in the teaching and learning of mathematics. It is assumed, with little evidence to support it, that there is likely to be transfer of learning from one problem situation to another if children are encouraged to seek alternative paths. Cockcroft (1982) supports this assertion by claiming that 'Not a great deal is yet known about the ways in which these [problem-solving] processes develop'. A major research study is required to shed light on this aspect of children's learning.

When the idea of open-ended problems entered the vocabulary of the mathematics educator, the known and recognizable goal which was part of previous definitions of 'problem' became redundant. 'Open-ended', when related to problems, means that the closed nature of a stated goal no longer exists in the statement of the problem, but is implicit. Consider the 'problem' quoted earlier:

> Investigate the addition of odd and even numbers.

Would you consider that this 'problem' is open-ended? How would you justify your decision?

The very narrow and limiting task-goal definition of 'problem' which has been part of mathematics curricula for so many years appears to have outlived its usefulness. It now becomes necessary to provide a definition of 'problem' that incorporates all the different facets of mathematical problems that children are likely to meet, including the idea of 'investigations'. In order to provoke discussion, I suggest that a problem is a situation that has interest and appeal to a child, who therefore wishes to explore the situation more fully in order to gain understanding of it. Goals arise naturally during the exploration and are determined not by the setter of the problem but by the child. The child in turn surveys the problem situation before exploring avenues of interest, following paths which may or may not lead to a satisfactory conclusion. As Ernest (1991) so succinctly puts it, 'the emphasis is on the exploration of an unknown land rather than a journey to a specified goal'.

WHAT IS AN INVESTIGATION?

It is difficult to determine when the word 'investigation' was first used in relation to a mathematical activity. What is certain is the extent of disagreement about answers to the questions 'What is an investigation?' and 'What is the difference between a problem and an investigation?'

Consider the 'problem' quoted earlier:

> Explore polygonal numbers.

Is this an investigation or a problem? What are your reasons for your decision?

After three decades, the stage has been reached in the development of problems and investigations in the curriculum which demands that mathematics educators decide what it is they are teaching. The misuse of the words 'problem' and 'investigation' is commonplace in mathematics classrooms. How is a child to make sense of any approach to learning mathematics when teachers in the same school cannot agree on the meaning of the two words? Even if the teachers have reached a consensus, their practice in the classroom may betray underlying differences. Fisher (1987) sums up the lack of unanimity in the profession by asserting that 'Mathematicians themselves are not in agreement as to the distinction between problems and investigations'.

Although the Cockcroft Report (1982) added emphasis to the importance of the role of problems in the mathematics classroom, there appears to be some confusion among mathematics educators as to the extent and clarity of Cockcroft's contribution. Costello (1991) suggests that the report 'spells out investigational work quite clearly, and gives investigation some emphasis by devoting several paragraphs to the idea'. In contrast, Evans (1987) says that although 'Cockcroft makes a distinction [between problems and investigations] . . . he doesn't say what it is'. Indeed, on closer inspection the Report has added to the confusion rather than clarifying the issue. In terms of what is often now referred to as 'investigational mathematics', as opposed to 'investigations', Cockcroft completely fudged the issue. Unfortunately, HMI (1985) claim that there is no clear distinction. Some amplification is provided, however, in the suggestion that 'in broad terms it is useful to think of problem-solving as being a convergent activity where pupils have to reach a solution to a defined problem, whereas investigative work should be seen as a more divergent activity'.

The distinction between 'problem' and 'investigation' cannot be dismissed lightly. There are serious implications for classroom practice if teachers are themselves uncertain as to the differences between problems and investigations. Children's mathematical behaviour and responses are frequently determined by the language used by a teacher, or which appears in a textbook. Solving problems which have a known goal requires a very different form of mathematical behaviour and inquiry from that demanded by an open problem or investigation where a goal is unstated.

In her short but informative book, *Mathematical Investigations in Your Classroom*, Pirie (1987) claimed that 'No fruitful service will be performed by indulging in the "investigation" versus "problem-solving" debate'. This is an evasion of a crucial issue which does a disservice to how and what we teach children. However, she goes on to describe investigations as presenting open situations which for children have no known outcomes. We should recognize that there may, in fact, be no known outcomes for the teacher either. Problems which have a set goal can be said to be closed and convergent, whereas investigations are usually thought of as being open and divergent. It is, however, possible to consider every mathematical situation to be a 'problem' in some form or other. As will be seen in the next section, there are those who in the categorization of problems include exercises.

So far, references to problems have included open-ended problems, open problems and investigations. The view that investigations are or are not problems differs from country to country, and within countries. If it is considered necessary to retain an independence for investigations, it is also necessary to show their relationships to problems. An attempt to do this is summarized in Figure 10.1. This claims that a pupil is presented with an investigation when the context is a situation which leads to a mathematical goal being chosen as an outcome of the exploration of the situation. The method of solution, if a solution exists, is also the choice of the pupil. Thus there are two types of investigation. The first, open-ended problems, involves pupils searching for a goal which they know is implicit in the problem statement. The second type, open problems, has two sub-types, (1) when there is no apparent goal until a pupil chooses to decide on one, and (2) when the goal is clear but the method is completely open. Thus,

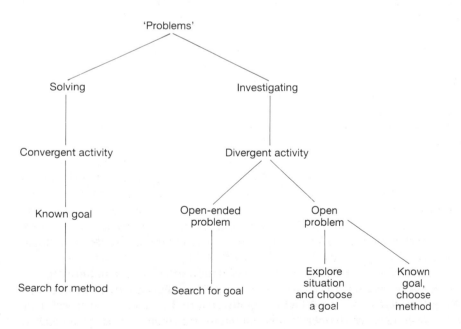

Figure 10.1 *Relationships between problems and investigations*

sorting numbers according to their factors is an example of an open-ended problem, as the criteria for sorting are not stated. Making a box, on the other hand, clearly specifies the goal but the method for so doing is left open. Finally, investigating triangles would constitute a problem with neither goal nor method made explicit. Now return to the seven 'problems' quoted earlier. Into what categories would you now place them? Can you justify your decisions and list your reasons?

It is important, here, to discuss the differentiation which can be made between an investigation and investigative working or investigative mathematics. The word 'investigation' is a noun and represents a category or type of problem, previously referred to as an 'open problem'. There is a danger that investigations are seen as ends in themselves and not as particular examples of a more general approach to learning. This more general approach involves teaching as much of the curriculum as possible in an investigative or exploratory way. Those in the mathematics teaching profession who advocate an investigative approach to learning the subject claim that it provides a fertile experience of the processes involved in mathematics and in mathematical thinking. Doing 'investigations' which are not integrated into the approach to the whole mathematics curriculum is known as the 'bolt-on' approach, and carries dangers which will be considered later.

CATEGORIES OF PROBLEMS

The distinction between closed and open problems is only one way in which problems can be categorized. This categorization is based on the nature of the outcome of a problem. Both of these dimensions of problems are useful in that they provide teachers with a way of ensuring that the curriculum activities which they devise for pupils are set in challenging and varied contexts. An alternative approach, which is used in the NC, classifies activities and hence problems associated with them as being practical, real-life, or investigations. This latter categorization, however, is somewhat confusing as it suggests that practical tasks or problems are not investigations; that real-life problems are necessarily different from practical activities. As teachers well know, there is considerable overlap between all three types. Many educationists outside Britain would claim that there are other kinds of problems which are in the curriculum of their country and which do not fit into any of the three categories of the NC. One very important type which has already been mentioned is 'word problems'.

Word problems have been criticized earlier, but they do make a significant contribution to the mathematics education of children in many countries throughout the world. They have also received vast attention from the research community, partly because, due to the difficulties that children encounter in their attempts to solve them, they provide fruitful areas for studying pupils' thinking. Word problems take two forms. The first is that which is referred to as 'story problems' set in words. These are familiar to every teacher of mathematics, as they are often to be found at the end of a set of algorithmic exercises and aim to apply the practised algorithm in a pseudo-real-life context. They are partly the reason why so many pupils have unhappy memories of their school experience of mathematics; they also provide teachers with a well-nigh impossible teaching challenge. The second form of word problem occurs when mathematical questions which would normally be set in a symbolic mode are expressed in words. Thus 8 − 5 is translated into 'What is 5 from 8?', or, even more challenging, 'What is the difference between 8 and 5?' Word problems in both guises have been in mathematics curricula since the earliest of textbooks. That they still remain in the mathematics curriculum of many countries is testimony to the sabre-tooth tiger aspect of the content of mathematics curricula; they should be extinct. What contribution do such problems make towards the mathematical development of children? There is little doubt that word problems are unlike any of those so far discussed and to solve them is as much a linguistic exercise as a mathematical one. They are also so artificially fabricated and lacking in pupil appeal that they are often 'counter-productive in developing analytical skills' (Chisko, 1985).

Teachers should always be asking themselves, 'Why should I use this task or problem with my pupils?' Only through being able to place a potential pupil problem experience in a classification system will the purpose of the task be immediately apparent. Butts (in Krulik and Reys, 1980) provides a simple, yet workable, partitioning of mathematical problems into five categories, as shown in the table (below).

Routine tasks	Non-routine tasks
Recognition exercises	Open search problems
Algorithmic exercises	Problem situations
Application exercises	

The non-routine tasks appear to be similar to those previously shown in Figure 10.1, but include reference to 'open search problems' rather than to 'open-ended' or 'open problems'. As we shall see later, they are not really similar. It may not be apparent why including 'exercises' within the ambit of 'problems' is of value. Many would claim that an exercise is not a problem. The application of the criteria described earlier in this chapter would support such a contention. However, exercises and problems are inextricably linked; problems as previously defined can neither be investigated nor solved without the knowledge and skills which have been perfected through the practice of exercises. The categories also suggest that what teachers may consider to be exercises, whose purpose is to consolidate knowledge and skills already acquired by a pupil, may begin life as problems. They also suggest opportunities to the creative teacher for devising problems and investigations which are not only more challenging than the equivalent exercise, but provide the pupil with practice in a highly motivating setting.

A typical algorithmic exercise which involves no exchange or carrying is:

$$\begin{array}{r} 3\ 6 \\ +\ 2\ 3 \\ \hline 5\ 9 \end{array}$$

This is easily turned into a problem, one with a unique solution, by the replacement of one of the digits with an unknown in the form of a letter. Below is a sequence of such problems, which begins with one that has a unique solution and gradually moves to an open-search problem.

$$\begin{array}{r} 4\ 8 \\ +\ 2\ A \\ \hline B\ 9 \end{array} \qquad \begin{array}{r} 2\ 5 \\ +\ 1\ D \\ \hline E\ 1 \end{array} \qquad \begin{array}{r} 2\ X \\ +\ 1\ X \\ \hline Y\ 2 \end{array}$$

$$
\begin{array}{r}
2\ 6 \\
+\ 3\ L \\
\hline
5\ M
\end{array}
\qquad
\begin{array}{r}
2\ 5 \\
+\ 1\ P \\
\hline
4\ Q
\end{array}
\qquad
\begin{array}{r}
3\ 7 \\
+\ T\ U \\
\hline
9\ V
\end{array}
$$

Many will recognize the world of cryptarithmetic emerging in these problems. This is more obvious with problems such as:

$$
\begin{array}{r}
A\ B \\
+\ B\ A \\
\hline
C\ D
\end{array}
$$

Children find this type of problem both challenging and rewarding, as it allows them freedom to use their own approach and thinking rather than replicate that taught to them by their teacher.

Butts (1980) defines his open-search type of problem as 'one that does not contain a strategy for solving the problem in its statement'. Thus the openness refers to the method of solution, not to the solution. He gives the following as an example:

> How many different triangles with integer sides can be drawn having a longest side (or sides) of length 5 cm? 6 cm? n cm? In each case, how many of the triangles are isosceles?

Some teachers would see this as an investigation, as it allows opportunity for a particular problem which has a solution leading to further problems and eventually into a possible generalization. However, there is always a solution to be found, whether numerical or algebraic. This is implied in the problem when the solver is asked, 'How many . . . ?' Such a position assumes that although investigations exist independently of problems, they can also arise from the solution of problems. Thus a distinction should be made between open-search problems which lead to investigations, and open problems or investigations which have their own separate existence. Again we have similar language used to describe the two different types of problem. The open-search problems give little indication, if any, of the mode of attack, while the open problems not only leave open the method, but also refrain from stating the goals, these being determined by the pupil in the course of investigating. It is, however, nearly always possible to restate an open-search problem in order to make it into an open problem or investigation. This is a very useful device for teachers who may wish to use textbook problems as the basis of investigations. The above problem can be restated to convert it into an investigation as follows:

> Investigate triangles which have integer sides and have the longest side (or sides) of length 5 cm.

Although one would anticipate that a pupil performing this investigation would at some time pose the question, 'How many such triangles are there?', it is not stated explicitly in the investigation. In a similar way, it would be expected of a pupil experienced in working with investigations that after exhausting the investigation as stated, the question, 'What would happen if . . . ?' would be posed. This would give rise to repeating the activity with 5 cm replaced by 6 cm, and later by other whole number lengths, eventually, if possible, generalizing for n cm.

The investigation is even more open than has been described so far, as a pupil may decide to move in a different direction by considering triangles with integer sides having the shortest side (or sides) of length of 5 cm. Later the pupil may wish to move away from triangles and investigate quadrilaterals having the same constraints. The generalization in this case, if one is possible, is in relation to polygons having integer sides and with a longest side 5 cm. Pupils who commence their mathematical education with closed problems move through the open-search problem phase into investigations, finally becoming problem posers, creators of their own problems.

Problem situations as described by Butts are very similar to the practical and real-life problems referred to in Ma1 of the NC. There are, however, degrees of openness which can be applied to these problems in the same way as to open problems. The following are three problems, all related to a trip to Alton Towers (a theme park), but differing in their openness:

1. If 38 pupils wish to go on a school trip to Alton Towers, how many coaches will be needed?

2. A school trip is being organized to Alton Towers. How many coaches will be needed?

3. Organize a school trip to Alton Towers.

The first two problems end with a question, although both require pupils to find further information and make decisions from choices. The final problem is an instruction to perform a task for which the minimum amount of information is given. Again, as with true investigations, this final type of practical or real-life problem should be the goal which teachers seek to attain with their pupils. Only when their pupils can approach such investigations and problems with confidence can teachers claim that the pupils have become independent learners through problem posing.

There is a major question related to problem situations or real problems, which should concern every teacher because of its implications for pupils: 'What does "real" mean?' Even if it is possible to establish a commonly agreed meaning, a more fundamental question then arises, namely, 'Is it possible in the classroom environment to create problems which children accept are "real" for them as a group or as individuals?' Lester (1989), when reviewing research into the use of mathematics in everyday situations with pupils, makes the point that 'problem-solving in the real world usually is sustained by the fact that the contexts in which the problems are embedded make sense to the problem solver'. The challenge to teachers lies in their ability to create real problems which are part of pupils' reality and not of that which an adult thinks should be or would wish to be part of pupils' reality. The three problems about a trip to Alton Towers illustrate this concern particularly well. Visiting Alton Towers may be of interest to pupils, but would they be sufficiently motivated to wish to calculate the cost of a trip that may never take place? The problem will not take on a sense of reality for pupils unless its purpose relates to a visit to Alton Towers which really will take place, and then the pupils feel a commitment to solving the problem. If the trip is merely a 'pretend' context set up for the sole purpose of practising mathematics, then pupils will go through the motions as they do so often with much of what passes as mathematics in classrooms.

The following criteria give an idea of the kind of problems which can be classified as 'real'. Every problem situation should attempt to satisfy as many as possible of the following:

1. The problem should be posed by pupils.
2. Pupils should wish to solve the problem for its intrinsic worth.
3. The problem should be meaningful for the pupils, not just for the teacher.
4. The problem should have relevance to pupils' everyday life in and out of school.
5. The method of solution should not be immediately apparent and there may be many solutions, each depending on the method chosen.

Teachers should be prepared for pupils tackling real problems with enthusiasm, but using little of the mathematics which they have been taught in the classroom. As Lester (1989) again stresses, 'in the everyday world people use mathematical procedures and thinking processes that are quite different from those learned in school'. These should not be rejected by teachers because they appear unrelated to what they have spent so much time teaching. Nor should pupils feel that the only kind of mathematics they are permitted to use when working on real problems is their 'school mathematics'. Lester has some reassurance for teachers with his claim that 'people's everyday mathematics often reflects a higher level of thinking than is typically expected or accomplished in schools'. He further claims that 'when exact calculations are needed and made, school-learned algorithms are often *not* used, and . . . the answers are almost always correct'. The challenge for teachers in using real problems is to take advantage of pupils' informal knowledge and skills, making them a basis for their future mathematical development, enabling them to apply mathematics to a much wider diversity of experiences than those in their limited and confined worlds.

PROCESSES, STRATEGIES AND SKILLS

Shuard (1986) asserts that 'Much of the recent interest in processes stems from Polya's work on mathematical problem-solving'. Reference to processes as essential features of investigations and problems abound in the literature. Some writers give examples of what they consider to be processes. Burton (1983) suggests that 'At the early childhood level, appropriate problems are easily structured around three important mathematical processes – classification, seriation, and patterning'. It is noticeable that Burton writes of 'mathematical processes', whereas Shuard could quite easily be referring to 'thinking processes'. The NCC (1989), either consciously or unconsciously, does not refer to the Statements of Attainment (SoA) in Ma1 as processes, yet this Attainment Target (AT) is now commonly referred to as 'the process AT'. The assumption here is that the SoA in Ma1 are concerned with the processes involved in using and applying mathematics in practical tasks, in problem solving and within mathematics. The NCC write only of 'the three strands of mathematical activity', these being (1) Applications, (2) Mathematical communication, and (3) Reasoning, logic and proof. The Inner London Education Authority (ILEA, 1990) further confuse the issue by introducing a further term, namely 'strategy', when they assert that 'The strategies used in mathematical work (the "processes" as they are often called) are outlined broadly in . . . (Ma1)'.

What then are 'processes' in relation to investigations and problems? Shuard (1986) claims that 'We *do* a process. Processes are actions or verbs.' In contrast, Shufelt and Smart (1983) view problem solving as one process: 'Problem-solving is a process, not a step-by-step procedure or an answer to be found; it is a journey, not a destination.'

It would appear that there may be many processes which make a contribution to an investigation or to the solving of a problem, yet the summation of these processes is itself considered to be a process. There is a conflict here which needs resolving if teachers are to be able to justify the inclusion of processes, as well as the traditional content, as essential elements of teaching and learning mathematics.

Backhouse *et al.* (1992) suggest that 'In general terms, a process can be thought of as a way of doing something, or a mode of action.' They are the means by which pupils put concepts, knowledge and skills to work. The nature of the 'doing' helps to classify processes; they are the 'ings' of the mathematics curriculum. In the classification which is discussed below, it becomes apparent that processes are broad in their capability. The generality of processes is their strength in that they are applicable to a whole variety of different problems and investigations. At the same time, from the point of view of a learner, processes are particularly difficult to work with consciously as there are so many of them. Not only do pupils need to 'learn' about processes, but they need to be able to choose at any given instant in the exploration of a problem which process is the most appropriate for the existing circumstances. This is very challenging for most children, and the ability to operate with a large choice requires vast experience of using processes in many and varied situations. This would not be serious if the introduction of processes to pupils was a gradual familiarization as part of a well-planned approach, organized over the total period of schooling. Each tier of school would need to recognize its role and know where to place the emphasis and on which processes. This may be the case in future in England and Wales, as the NC through Ma1 is making some contribution towards achieving this goal. However, there is much uncertainty whether processes can be taught, or whether they are only assimilated into a pupil's repertoire through usage over a long period of time. There is no established hierarchy of processes; it may be that one does not exist. Until more is known about how pupils learn processes and how and when to apply them, teachers are clutching at straws in their attempts to assist pupils to explore problems and investigations successfully. But it is only through the efforts of teachers, however much they are operating in an unknown area, that sufficient experiences will be collected to enable mathematics educators to move progressively towards greater understanding of this relatively new approach to learning.

Not all processes used in attempting investigations and problems are mathematical. Indeed, most processes are independent of mathematical content and are useful in problems associated with other subject areas and disciplines. Figure 10.2 illustrates a categorization of these subject-independent processes in relation to the mathematics-specific processes which are discussed later. The processes of communication, reasoning and recording will be familiar to most teachers who have used investigations and problems in their classroom. Operational processes may also have a ring of familiarity, but not under this umbrella. They are processes which pupils use to 'operate' with or on data. They include collecting data at the commencement of an investigation, sorting data to bring some semblance of sense to what may appear to be a random and chaotic collection, matching and comparing data to explore samenesses and differences between items of data, sequencing and later ordering data to explore relationships based upon rules, changing data and searching for the effect of chosen transformations, and finally combining items of data to produce new items or ones which already exist within the collection. The use of some of these operational processes will be illustrated in an example later in this chapter.

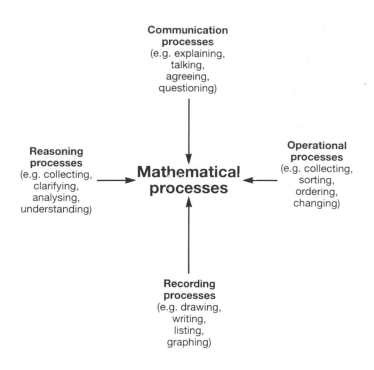

Figure 10.2   *Processes involved in exploring mathematical investigations and problems*

As communication, reasoning, operational and recording processes are applicable across the curriculum, there is a strong case to be made for close liaison between subject areas in a school to combine forces to develop pupils' abilities in operating with these general processes. At the present time most subject teachers work independently, developing their own approach to the exploration of problems which they see as unique to their specialism. The approach may indeed only be pertinent to their own subject, but the processes which are used to implement the approach are more general and less subject-specific. There are, of course, processes which are particular to each subject discipline, although it is often difficult to place a subject barrier around processes. Figure 10.3 lists some of those processes which mathematical educators might claim are uniquely appropriate to mathematics. It also attempts to show how the processes may be linked in an order of applicability when investigating and solving problems.

Knowledge of the relationships which exist between mathematical processes, as the use of one leads naturally into another, should be second nature to teachers of mathematics. It should be the aim of a problem-centred mathematics curriculum to move pupils also towards this state. As pupils use and apply processes in new situations, they develop an expertise in recognizing when to call upon a certain process as opposed to others, and in operating with it successfully. When this occurs a process is often referred to as a skill; they become skilful in its appropriateness and applicability.

The list of mathematical processes in Figure 10.3 is not intended to be exhaustive. One particularly important omission is that of 'symbolizing', a process perhaps not unique

Processes 'unique' to mathematics

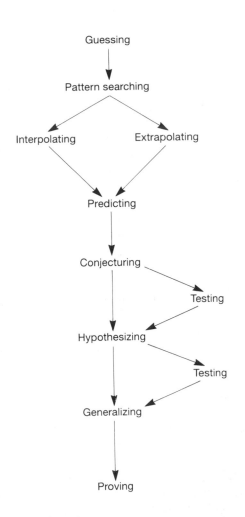

Figure 10.3   *Processes 'unique' to mathematics*

to mathematics, but one which provides the basic building blocks on which the whole edifice of mathematics is constructed. Unfortunately, pupils are frequently presented with ready-made, well-established and commonly agreed forms of symbols when being taught a new topic. Seldom are pupils given the opportunity to create their own symbols to express the ideas which make up the topic. Symbolizing is a process which all pupils should recognize has occurred in the making of mathematics throughout its history and will continue wherever mathematics is being created. This kind of experience should be given to every pupil. Many opportunities for such experiences arise when pupils are exploring problems or working with investigations. The need to record and communicate ideas in a succinct, precise and unambiguous way is an essential element of working with investigations and problems.

Polya (1945), in *How To Solve It*, described his now famous four-step plan for problem-solving. Essentially the steps are:

1. understanding the problem;
2. devising a plan;
3. carrying out the plan;
4. looking back.

The four steps provide a framework in which the processes used to solve a problem are chosen and ordered, but their applicability to the present approach to teaching investigations and problems is questionable. Frequently, the understanding of a problem only emerges slowly as it is explored. If the problem is known to have a solution, then Polya's first step requires the solver to recognize the existence of a solution. However, if the problem is open with respect to possible solutions and the method of attack, it is difficult for a pupil to develop an understanding of the problem without first attempting to explore whatever appears appropriate. Asking pupils to try to understand a problem before they begin to work on it demands an ability to reflect consciously upon previous and relevant experiences. This is a mature and sophisticated approach which most pupils find impossible to perform with any degree of success.

Polya's four steps are easily memorized and recalled by pupils; their implementation is of a much higher order of difficulty. Devising a plan demands some knowledge and understanding of the mathematical structure of the problem; yet exploring the problem aims to develop some appreciation of the structure. Here lies a barrier which results in a sense of frustration for pupils: they wish to 'play around' with ideas, but they cannot do so until they have constructed a plan. Polya's work has made a valuable contribution to the development of problem solving in schools, but problems as perceived by Polya are not the kind of problems and investigations which pupils now work on in classrooms.

HMI (1985) did not help the debate about processes in the mathematics curriculum when they listed what they called eight 'general strategies'. These were:

- ability to estimate;
- ability to approximate;
- trial and error methods;
- simplifying difficult tasks;
- looking for pattern;
- reasoning;
- making and testing hypotheses;
- proving and disproving.

Shuard (1986) considers that 'These strategies are at very different levels of generality, and some seem nearer to content than they are to process'. In fact, it could be claimed that they are not strategies. If anything they are processes, as they are what teachers expect pupils will do with concepts, knowledge and skills when exploring a problem or an investigation. What then is a 'strategy'? A strategy is a collection of processes placed in an order in which they may be used. Strategies can be at many levels of complexity. Pupils who decide that their first action with problem data will be to sort it have operated a strategy. They have chosen from all the processes available to them one particular process which they see as appropriate. Many pupils, when experiencing investigations and problems for the first time, adopt the most simplistic of strategies; after working with one process, they ask either of the teacher or of themselves 'What shall we do next?' Only after much experience of a variety of problems which give opportunity for using many different processes are pupils able to string together a planned sequence of processes forming a more complex strategy. Observing the nature and number of processes which pupils plan to use, and in which order, assists teachers in evaluating the development of pupils in exploring problems and investigations.

If teachers are clear about the distinction to be made between process and strategy, then their use of the words with pupils will carry clarity and an unambiguous meaning. Pupils will not be confused as to a teacher's intentions and will themselves be secure in their ability to operate with processes and plan workable strategies leading to successful conclusions. This can only lead to pupils tackling problems and investigations with increased confidence.

THE ROLE OF THE TEACHER

The teacher's role in a problem-centred classroom diverges from the traditional model (see Chapter 3). Pirie (1987) discusses the role under seven headings, namely:

1. the role of instigator;
2. the role of enabler;
3. the role of facilitator;
4. the role of listener;
5. the role of questioner;
6. the role of positive evaluator;
7. the role of observer.

Some teachers may claim that they already perform these roles in their present classroom. This may be true, but when the exploration of problems is taking place with a class organized in groups and working in a co-operative mode, the nature of the roles is very different from when operating in a more formal classroom organization.

The first change which teachers experience when introducing exploration of problems involves a loosening of the reins of control. Pupils begin to make their own decisions, some of which a teacher may not initially find acceptable. A teacher can develop a sense of insecurity as control passes to the pupils. This quickly passes as pupils use their new-found freedom to work closely together, co-operating in groups. Teachers need to develop their skills of observation, standing back from the fray, yet sensing when it is appropriate to make a suggestion or an observation on the work which is being discussed. To allow pupils to disagree is foreign to the way many teachers have operated a conventional curriculum in a formal classroom setting. But the essence of pupils' ability to develop the concept of proof lies in discussion, debate and disagreement. To argue one's case and to explain one's thinking to others is an early phase in the process of proving.

Teachers working with investigations and open problems for the first time will experience the urge to 'tell'. Resist the temptation; become a listener, not a teller! This is much easier to write than it is to achieve. Allowing pupils to spend time moving towards what a teacher knows will eventually lead to a dead end can be both frustrating and annoying. What must be recognized by teachers is that they only know that it will lead in to a dead end because they have been along that or similar paths. Pupils also need to have the same experience (see Chapter 3).

In a problem-centred classroom, pupils are encouraged to work co-operatively where appropriate. One of the advantages of working in this way for pupils is the discussion of the mathematical problem with one another. They share thoughts and ideas; they reason aloud; they express themselves in a coherent way, marshalling their thinking to present an argument or suggestion (see Chapter 7). Pupils are also drawn into working on problems in an active way as opposed to listening to the teacher. This can only occur if a teacher is prepared to allow it to happen and not intrude at inopportune times. This is not to say that a teacher should never make a contribution to pupils when they are working together. There will often be the need to respond to difficulties, which will occur frequently. On occasions a group will seem to have come to what they consider to be the end of the problem. The group members may need a stimulus from the teacher to enable them to 'recharge their batteries' and move forward to further explorations.

Pupils do not readily co-ordinate their efforts, and teacher input in such circumstances leads to greater efficiency of group operation. When a teacher does this, however, it should be made apparent what has been done and why. In this way pupils will develop the skill of organizing their efforts in future problem situations. It is to be expected that the unexpected will arise when pupils are exploring problems. The teacher has a significant role in responding in such circumstances. Pupils frequently need reassurance when the unexpected befalls them. They are easily deterred from continuing and it may need the teacher to work closely with them until they are confident enough to move on again without assistance.

Teachers who wish to adopt a problem-centred approach to their teaching should, themselves, have encountered the exploration of problems and investigations. It is essential to have experienced the frustrations and the joys which accompany such an approach to the study of mathematics. Having explored problems which are then given to pupils has disadvantages. The explorer now has a knowledge of avenues of inquiry which produced successful outcomes. Knowing what directions are the most fruitful to pursue can lead unwittingly to suggesting what and how pupils should explore in a particular problem. This temptation will also occur after problems have been used with one class and are then presented to another class the following year. Teachers should accept that the exploration of a problem or investigation is never complete. New groups of pupils are highly likely to produce similar work, but they could arrive at an idea which has never arisen with pupils in previous years.

INTENTIONS AND MATHEMATICAL BELIEFS

It would be misconceived if a teacher were to introduce a problem-centred approach into the classroom unless fully convinced of its value. The positive attitude and enthusiasm which this conviction brings to the atmosphere of a classroom is contagious and, as Fisher (1987) says, 'can have a profound influence on pupils' approach to mathematics'. Lester (1980) goes further, claiming that 'It also seems obvious that the likelihood of improved problem-solving performance is increased if students see good problem-solving behaviour exhibited by their teacher.' There are thus two aspects of teaching raised by Fisher and Lester. The first is to do with an approach to teaching and learning of mathematics, the second with the possible outcomes in terms of increased attainment. That both of these are a consequence of adopting a problem-centred curriculum has yet to be conclusively established. However, lack of evidence should not deter an enthusiastic teacher who holds the sincere belief that this approach is the right one.

It is *teachers* who develop the classroom environments in which the learning of mathematics flourishes or dies. A problem-centred environment cannot be created if a teacher's belief about mathematics and the nature of school mathematics is contrary to a problem-based approach to teaching. Although a teacher may hold the view that mathematics is about problem solving or exploration, there is no contradiction in perceiving school mathematics as content-oriented, with the aim of developing basic skills for application at some later time. Teachers who are considering adopting a problem-focused curriculum should carefully search their belief about the nature of mathematics (see Chapter 1) and how it relates to school mathematics. It would be wrong to proceed on the basis that mathematics and school mathematics are one and the same, even though we may believe that they should be.

Unfortunately, there is growing evidence that teachers, partly due to the pressures of GCSE, are interpreting problems in a very narrow way, viewing them as just another topic in the syllabus. Ernest (1991) claims that 'Several studies have revealed teachers who espoused a problem-solving approach to mathematics teaching . . . but whose practices revolved around an expository, transmission model of teaching enriched by the addition of problems.' The danger lies in teachers falling into line with the prevailing fashion, not basing their practice on a sincerely held creed.

Introducing problems and investigations into the curriculum can encounter resistance from pupils. Just as teachers hold opinions, so do pupils. These will not have developed overnight: 'They develop slowly, over a long period of mathematical encounters and experiences. Most students' primary source of mathematical experiences is probably the mathematics classroom' (Frank, 1988). Pupils whose experience of mathematics is one of providing the teacher with replicas of what was explained on the chalkboard do not welcome an approach which demands that they do some thinking for themselves. All too readily the cry emerges, 'What do you want us to do?', or even more worrying, 'Tell us what to do and we will do it'. Pupils learn attitudes as well as mathematical content from their classroom experiences.

The solution to overcoming the negative reaction of pupils to open problems is never to allow such attitudes to develop. This can only be achieved by pupils experiencing problems and investigations from the moment they enter the reception class or the first standard or grade. If the approach is continued through subsequent classes, standards or grades, pupils' perceptions of mathematics will be quite different from what is found to be the case at the present time. If, however, pupils are already conditioned to a diet of chalk, talk and exercises, then successful implementation of a remodelled curriculum requires changes in pupils' attitudes to mathematics, and eventually in their conceptions of the subject. Pupils who perceive mathematics as a collection of exercises to prepare them for tests and examinations do not readily accept that open problems and investigations constitute real mathematics. They do not believe that the activities which they are asked to do are appropriate, because of their view of mathematics, and they consequently reject the tasks. Teachers wishing to change their approach should recognize that pupils have firmly held beliefs, and that these can only be changed with careful handling and over a considerable period of time. Do not expect pupils to run before they can walk.

To this end it is helpful to commence the change with short open-search problems for which a unique solution exists. The method of reaching the solution should not be apparent, requiring some input from the pupils. Success should be achieved in a reasonable time. Short-term goals at this stage are essential rewards for pupils accustomed to similar acknowledgement for effort when completing exercises. An example which pupils enjoy and which has proved successful at an early stage of the change to a problem-focused approach is that of number pyramids (see Figure 10.4). At a later stage the same idea can be used to introduce the concept of negative number or to investigate when an infinity of solutions is possible (see Figure 10.5). Number pyramids have the potential to form investigations with limitless possibilities. They have been used successfully by pupils for their GCSE coursework assessment, yet can also be used with young children who are practising simple addition.

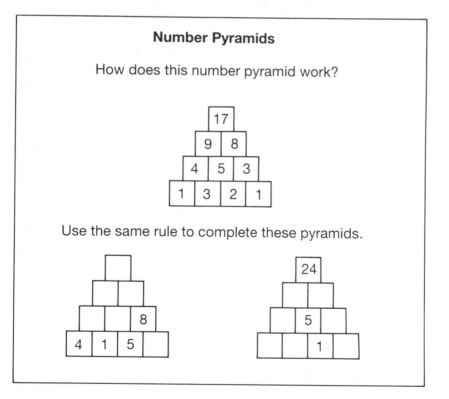

Figure 10.4 *Number pyramids*

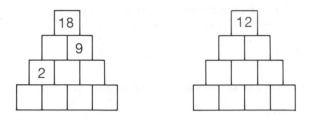

Figure 10.5 *More number pyramids*

AN INVESTIGATIVE APPROACH TO LEARNING MATHEMATICS

Shuard (1986) recognized that investigations are not ends in themselves. Early experience of observing investigations in classrooms led her to warn that 'There is, however, a danger that investigations will be seen as an isolated part of the mathematics curriculum . . . which does not spill over into the remainder of the mathematics curriculum.' The bolt-on approach to including problems and investigations as part of a mathematics curriculum is doomed to failure. It provides pupils with conflicting messages about mathematics and, what is more worrying, about how mathematics is learned. It is not uncommon for a teacher who has adopted this approach to start a lesson by saying, 'Today we are going to do an investigation', only to be met by the response from the class, 'Not another investigation. We did one last week.' A problem-centred or investigative approach to learning mathematics should be perceived by pupils as an ideal way to learn the subject. An investigative teaching style is a pedagogical approach to the mathematics curriculum, not just something which occurs when the routine of the normal curriculum becomes dreary and tiresome.

The essence of an investigative approach is the application of communication, reasoning, operational and recording processes to a study of the core topics which make up the content of a mathematics curriculum. An example will be used to illustrate how these processes contribute to learning mathematics. The example, addition pairs (see Figure 10.6), has been used successfully with a wide range of ages and abilities. It is presented here on an activity sheet, but its mode and format of presentation should be chosen to match the needs of pupils. It is helpful to provide pupils with small, rectangular pieces of paper on which to record each addition pair that they make. Writing each pair on the same sheet of paper creates data, namely the addition pairs, which are static, unable to be moved around. This moving around of 'dynamic' data encourages pupils to put their thinking and their conjectures to the test, and if necessary to modify their first thoughts. Once the dynamic process has produced an acceptable classification, the data may be recorded in static form.

The initial activity is the collecting of data, the addition pairs. Pupils do this for themselves either individually or co-operatively, pooling their pairs. Working in groups encourages them to see whether they have produced any which are the same or whether they are all different, an early aspect of the operational process of matching. Usually, there is immediate disagreement, as some children use '0', while others claim it should not be allowed. It is necessary to agree on the rules of the problem space, or what could be called the constraints of the system. Other pupils ask whether the two pairs 1 + 2 and

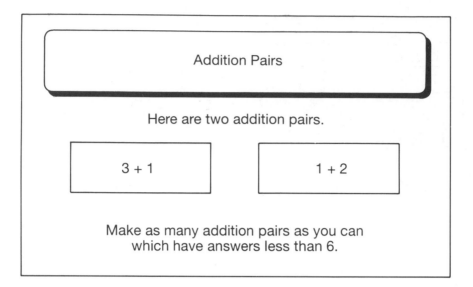

Figure 10.6   *Additions pairs*

2 + 1 are the same or different. Once again it becomes necessary to discuss this with everyone in order to arrive at an agreed 'definition'. If you try this with your classes you may also find that some pupils will write numbers involving fractions. Again, everyone has to agree whether to permit fractions or not. It is advisable to restrict young children to whole numbers. In this way the quantity of data does not overwhelm them; there are not too many to make the activity unmanageable. The end result may be all possible pairs 1 + 2, 3 + 2, 1 + 1, 1 + 3, 2 + 1, 4 + 1, 2 + 2, 1 + 4, 2 + 3 and 3 + 1. I have deliberately not listed the pairs in a sequence or pattern as few children adopt such an approach when collecting data unless they have had considerable experience of doing so.

The activity contains a particularly important expression, 'Make as many addition pairs as you can'. The children are not asked, 'How many addition pairs are there?' It is the act of making the pairs which is mathematically important. How many there are is a consequence of the making of the pairs and not the sole objective. To ask 'How many?' is a threatening question for many children, who have no obvious means of finding out and who may in consequence resort to wild guesses. Making pairs, producing data, provides very short-term goals, with pupils achieving success every time they make another pair. Sorting, sequencing and ordering data are operational processes which encourage children to consider each addition pair carefully in relation to a given criterion. Initially children need assistance in deciding on ways of sorting, but this stage soon passes as they quickly challenge each other with complex criteria for sorting. Sorting can take place on grids with space provided for attributes or properties which determine the sort. Figure 10.7 shows an example which children readily suggest. Children appear to have a natural ability to sort using pattern. When sorting in the way shown in Figure 10.7, the vast majority of children end up with the pairs ordered in each set. They then search for the patterns, a mathematical process applied to the data. The

1 is on the left	2 is on the left	3 is on the left	4 is on the left
1 + 1	2 + 1	3 + 1	4 + 1
1 + 2	2 + 2	3 + 2	
1 + 3	2 + 3		
1 + 4			
4	3	2	1

Figure 10.7 *Sorting of addition pairs*

sorting grid can have an extra section along the bottom in which a number can be placed to indicate how many pairs have been found in each set. This also produces a pattern of numbers when all possible pairs have been sorted which may be recognized as the counting numbers.

There are many other activities of this nature using operational processes which assist children to develop an awareness of the structure of this 'simple' mathematical system. The system can be extended in a number of different directions. Relationships between the different systems can then be explored. New problems arise continually, and are created by the children; for example, 'What happens if triples of numbers are allowed? quadruples? . . . $n$tples?', or 'How many pairs? triples? quadruples? . . . will there be if the sum remains less than 6?' The children are now involved in recording new data and then exploring relationships between the sets and looking for a possible generalization (see Figure 10.8).

Numbers in the sum	The number of additions
2 (pairs)	10
3 (triples)	?
4 (quadruples)	?
n ($n$tples)	?

Figure 10.8 *Exploring addition* n*tples*

CONCLUSION

It has only been possible to give a flavour of how an investigative approach can be adopted to 'teach' a topic in the curriculum. If you try this with your pupils, remember to be an asker and a listener, not a teller. Ask questions when they will assist pupils to move on. Listen to what they have to suggest; you will be surprised how quickly they can show responsibility for their own learning, posing their own problems which they then proceed to solve. Whether it is possible, or even desirable, to use an investigative approach to teach every topic in the mathematics curriculum is a question which most schools are far from asking. The first task for any school is to make a start, however small it may be and however slowly change takes place. But before a school decides to adopt a problem-centred or an investigative approach (are these the same or different?), teachers should examine their views on mathematics, and only if they feel their perceptions of the subject match the philosophy of an investigative approach should they then proceed.

REFERENCES

ATCDE (1967) *Teaching Mathematics: Main Courses in Mathematics in Colleges of Education.* Cambridge: Cambridge University Press.

Backhouse, J., Haggarty, L., Pirie, S. and Stratton, J. (1992) *Improving the Learning of Mathematics.* London: Cassell.

Burton, G. M. (1983) Problem solving: it's never too early to start. In G. Shufelt and J. R. Smart (eds), *The Agenda in Action.* Reston, VA: NCTM.

Chisko, A. M. (1985) Developmental math: problem solving and survival. *Mathematics Teacher* **78**(8), 592–6.

Cockcroft, W. H. (1982) *Mathematics Counts.* London: HMSO.

Costello, J. (1991) *Teaching and Learning Mathematics 11–16.* London: Routledge.

Ernest, P. R. (1991) *The Philosophy of Mathematics Education.* London: Falmer Press.

Evans, J. (1987) Investigations – the state of the art. *Mathematics in School* **16**(1), 27–30.

Fisher, R. (1987) *Problem Solving in the Primary Schools.* Oxford: Blackwell.

Frank, M. L. (1988) Problem solving and mathematical beliefs. *Arithmetic Teacher* **35**(5), 32–4.

HMI (1985) *Mathematics from 5 to 16.* London: HMSO.

ILEA (1990) *Mathematics in Primary School (Part 1): Children and Mathematics.* Sidcup: Harcourt Brace Jovanovich.

Krulik, S. and Reys, R. E. (eds) (1980) *Problem Solving in School Mathematics.* Reston, VA: NCTM.

Lester, F. K. (1980) Problem solving: is it a problem? In M. M. Lindquist (ed.), *Selected Issues in Mathematics Education.* Reston, VA: NCTM.

Lester, F. K. (1989) Mathematical problem solving in and out of school. *Arithmetic Teacher* **37**(3), 33–5.

NCC (1989) *Mathematics Non-statutory Guidance.* York: NCC.

NCTM (1980) *An Agenda for Action: Recommendations for School Mathematics of the 1980s.* Reston, VA: NCTM.

NCTM (1989) *Curriculum and Evaluation Standards for School Mathematics.* Reston, VA: NCTM.

Pirie, S. (1987) *Mathematical Investigations in Your Classroom.* Basingstoke: Macmillan.

Polya, G. (1945) *How To Solve It.* Princeton, NJ: Princeton University Press.

Reys, R. R., Suydam, M. N. and Lindquist, M. M. (1984) *Helping Children Learn Mathematics*. Englewood Cliffs, NJ: Prentice-Hall.
Shuard, H. (1986) *Primary Mathematics Today and Tomorrow*. York: Longman.
Shufelt, G. and Smart, J. R. (eds) (1983) *The Agenda in Action*. Reston, VA: NCTM.
Wolf, A. (1990) Testing investigations. In P. Dowling and R. Noss (eds), *Mathematics Versus the National Curriculum*. London: Falmer Press.

Chapter 11

Integrating Mathematics into the Wider Curriculum

Tom Roper

INTRODUCTION

Mathematics is widely perceived as being 'useful'; in the 'real' world, in everyday life, in one's present or future career and in the study of other subjects. This perception is often translated into reasons and/or aims for the teaching of mathematics (see Chapter 1). It is a perception encouraged by teachers and parents as a means of motivating pupils and one which pupils frequently accept, implicitly if not explicitly, in terms of the need to obtain certain prerequisites for future careers and educational opportunities. As a result, mathematics occupies a central place within the mainstream curriculum of all countries, both in its own right as a separate subject and as a fundamental part of an increasing number of other subjects.

However, these two aspects of mathematics seldom seem to meet within the wider school curriculum. Much lip-service is paid to the idea of integrating mathematics into the wider curriculum, but little of practical use to the student or the teacher would seem to take place. It is pertinent to ask why this is so. Is mathematics at all useful? In what ways is it perceived to be useful? How do these perceptions affect the way in which mathematics is viewed and taught?

THE USEFULNESS OF MATHEMATICS

The fact that we live in a society increasingly dependent upon mathematics as a way of representing, communicating and predicting events, information and future trends can be simply demonstrated by considering the contents listings of three books which discuss the applications of mathematics. The volumes, Sutton (1954), Holt and Majoram (1973) and Howson and McClone (1983), represent a time span of 27 years and each one was designed to be read by students of A-level mathematics at the time it was written. However, the contents pages and a glance at both the situations and the mathematics being used show remarkable changes. Sutton (1954) has applications of mathematics in fields which are based in physics, engineering and military traditions. The mathematics being used is therefore largely Newtonian mechanics and the theory of differential equations. The one exception is the use of statistics in weather forecasting. Holt and Marjoram (1973) illustrate the beginnings of a number of trends: first, the application of a wider variety of mathematical models, as opposed to the almost exclusive use of the Newtonian model; second, applications in new subject areas in addition to the physical sciences, including geography and environmental studies; third, the wide use of a range of models, including the Newtonian model, in all subjects. Howson and McClone (1983) continue to reflect the trends noted above, with the first trend showing an increasing use of statistical and discrete mathematics models as opposed to the continuous, and the second trend showing a widening of the fields of application which impinge more directly upon the daily life of the average citizen, for example in nursing and even in legal matters.

The use of mathematics in the form of mathematical models in such a wide variety of academic and practical situations suggests that mathematics is having an increasingly direct effect upon our daily lives. In the past it might perhaps have been possible to argue that mathematics was the tool of the scientist and that, as such, the effects at the immediate, personal level would not be directly discernible. This position is now even more difficult to sustain, if it ever was sustainable. Examples of the use of mathematics in the courtroom will illustrate the point. Evidence based upon the Poisson model of probability is now acceptable in a court of law in cases concerning parking meters and repeated parking offences (see Howson and McClone, 1983); Newtonian mechanics forms the basis of detailed police accident investigations; technology and mathematics combine to measure our speed on the motorway either directly via hand-held 'guns' or indirectly from video film. It is indeed a sobering thought that the mathematics we may treat so abstractly in the classroom could now come to our aid or otherwise in the courtroom.

Employment may also be subject to radical changes through the use or introduction of mathematics into specific situations or processes (see Fitzgerald *et al.*, 1981; Fitzgerald, 1985). People may find it difficult, if not impossible, to come to terms with some of these changes. One of the most striking developments is the use of discrete rather than continuous mathematical methods (see Chapter 5). Dr R. V. Peacock, discussing the case for the introduction of discrete mathematics into the mathematics curriculum for 16–19-year-olds at the HMI Llandudno Conference of February 1989, indicated the importance of discrete methods in industry and management, especially in solving organizational problems. In doing so he related the experience of a distribution manager who was unable to accept the practical results of a mathematical analysis of the distribution of goods from a factory. The manager apparently stopped every truck leaving the factory to photograph the load-space of the truck in order to show that many of them were not full. However, a full load was not an essential criterion of the cheaper, quicker and more efficient distribution service provided by the discrete mathematics model used.

This rapidly growing use of mathematics in an increasing variety of situations, academic, practical and social, with old models popping up in new places and vice versa, is reflected in many school-based subjects which have imported mathematics to a greater or lesser degree. Physics, and to a lesser extent chemistry, have always been major users of mathematics, as a provider of both tools and models. At the highest levels of theoretical physics, the subject is entirely mathematical, with experiment and observation acting to confirm or deny predictions from the model and pointing the way to its development or the introduction of a new one. Examples of other subjects which have increased their usage of mathematics, both in kind and degree, are geography, history, economics, sociology and biology.

Indeed, such has been the drive to import tools and models from mathematics into other subjects, coupled with the recognition of the importance of the idea of the mathematical model, that several British A-level syllabuses in mathematics have incorporated the model-building process. Examples include the School Mathematics Project 16–19 (SMP), the Mathematics in Education and Industry Project (MEI) and the Nuffield and Wessex projects. To a large extent these new syllabuses have been influenced by the 'modelling philosophy' of the Mechanics in Action Project (MAP), the model building featuring mainly within the mechanics modules or texts. However, the incorporation of the model-building process into an A-level standard syllabus is not a recent idea. The Schools Council Sixth Form Mathematics Curriculum Project, under the direction of C. P. Ormell, made the philosophy of C. S. Peirce central to the project (see Ormell, 1972, 1975, 1977; Peirce, 1956). This resulted in the whole of the project being based upon the idea of mathematical modelling, with the motivation for the areas of mathematics to be studied arising directly out of practical problems to be solved. The O-level Additional Paper which was offered through the University of London School Examinations Board as a means of examining the course is no longer available. The textbooks will, however, repay study, since they represent a unique and original way of motivating the teaching and learning of mathematics (see Mathematics Applicable, 1975, for full details).

The application or use of mathematics, as a tool and as a provider of models, has so far been discussed in a fairly undefined way with the assumption of shared meanings for the words and phrases used. These issues will be returned to later. However, having seen the increasing usefulness of mathematics within a wide range of subjects, we must turn our attention to mathematics across the curriculum. The difficulty of establishing mathematics across the curriculum as a practical reality cannot be over-emphasized. Despite many statements about the necessity of so doing in order to make mathematics a more integrated part of general education, little has been achieved to bring it about. In England and Wales, for example, a National Curriculum (NC) has been introduced which seeks to give emphasis to the 'using and applying of mathematics' by giving one of the five Attainment Targets (ATs) in mathematics, Ma1, this particular title. A consideration of the official statements and assistance given to the teacher in respect of establishing mathematics across the curriculum within the context of this initiative provides a useful case study.

THE VIEWPOINT OF THE NATIONAL CURRICULUM OF ENGLAND AND WALES

In its opening chapter, the Cockcroft Report (Cockcroft, 1982) sought immediately to address the question, 'Why teach mathematics?' It put forward three reasons, dismissed a fourth and acknowledged the existence of others but declined to list them. The chosen three were as follows:

> . . . foremost among them is the fact that mathematics can be used as a powerful means of communication – to represent, to explain and to predict . . . A second important reason for teaching mathematics must be its importance and usefulness in many other fields . . . the inherent interest of mathematics and the appeal which it can have for many children and adults provide yet another reason for teaching mathematics in schools.

There is a strong utilitarian thrust within the first two of these reasons; the pre-eminent reason, it is argued, is the basis of a wide variety of 'perceptions about the usefulness of mathematics' (Cockcroft, 1982). However, the language of 'communication', 'represent', and the like does not suggest a crude utilitarianism, of reaching for mathematics as though it were a tool, a hammer say. It implies the more sophisticated usage of mathematics as modelling to represent and predict.

In HMI (1985), no reasons are given for teaching mathematics but ten aims are advanced (see Chapter 1). The first two aims, 'Mathematics as an essential means of communication' and 'Mathematics as a powerful tool', would appear to reflect the Cockcroft perspective above. However, the explanatory paragraphs associated with the second aim seem to have the cruder point of view:

> A tool enables things to be done which it might otherwise be impossible or difficult to do, or to do as well. Mathematics is such a tool. Many instances arise in the school curriculum, in working life and in society generally where mathematics is a tool to be used in a variety of ways. Viewed from this perspective it is not the mathematics itself but the result obtained which is the important thing.

Again, later in the same paragraph,

> Skills such as . . . are not important ends in themselves and only become so as they are embedded in purposeful activities.

Either the mathematics to be used in the situations is, in some sense, thought of as being obvious in the same way that one would choose a screwdriver in order to fix a screw rather than a hammer, or the whole complexity of what mathematics might be appropriate to the situations being considered is being totally neglected. The examples used, for example 'checking a shopping bill', seem to suggest the former and hence the cruder interpretation of utility.

The National Curriculum Working Party Document (DES, 1988), in several places, for example paragraphs 3.15, 3.19, 3.22, 3.23, clearly espouses a utilitarian view concerning the aims of teaching mathematics. Indeed Noss (1990) expresses the view of the Working Party as being this: 'mathematics is a set of tools: its application is in the solution of problems and the mechanism is in the selection of the appropriate tool.' Noss views the NC in mathematics as being based upon a utilitarian perspective which fits well with the tool analogy rather than the modelling framework indicated by Cockcroft. Noss also suggests evidence for an underlying theme of social control through particular interventions by the then Secretary of State for Education. However, this aspect of Noss's critique does not concern us here.

The 1989 version of the NC in mathematics (DES, 1989) carries forward the recommendations of the working party almost unchanged. Certainly there is no restatement of philosophy. The revised NC in mathematics of 1991 (DES, 1991) again offers no restatement of philosophy. Thus a part of the rationale behind the existing NC must still be the somewhat crude utilitarian view of mathematics indicated above. However, non-statutory guidance (NCC, 1989, 1991) has been issued for the original curriculum and the revised version. The latter directs the reader to the former, having basically nothing to add to the advice therein offered. Within the former document, one of six sections is devoted to Ma1, 'Using and applying mathematics', and another section, section F, to 'Mathematics in the curriculum', which attempts to give reasons for developing and planning cross-curricular work based upon:

> efficient use of time through cross-curricular work . . . Mathematics pervades many areas of the curriculum and mathematical activity can contribute significantly to the development of more general skills such as communicating, reasoning and problem-solving. . . . Similar examples [of pupils needing mathematics to help them understand and to communicate] can be found throughout the curriculum. . . . Mathematics is a powerful tool with great relevance to the real world. For this to be appreciated by pupils they must have direct experience of using mathematics in a wide range of contexts throughout the curriculum.

From the above one might conclude that perhaps here was an attempt to redress the balance of the Working Party report and the assessment-driven nature of the mathematics curriculum as published (see Chapter 4). Within Ma1, 'Using and applying mathematics', the process elements of mathematics are established as part of the curriculum, and the wider curriculum is seen as a source of fields for the use of content and the practice of process, strong words of encouragement being given to teachers to look to the wider curriculum.

From the reading of official documents, two major points can be observed. First, although there are strong words concerning the integration of mathematics into the wider curriculum and the relationship between mathematics and other subjects, there is little, if any, practical help offered and few exemplars given. The National Curriculum Council (NCC) Consultation Report, Mathematics (NCC, 1988), states, 'Council advises that cross-curricular aspects will be more fully worked out when other core and foundation subjects are in place. There will need to be further non-statutory guidance on cross-curricular matters.' The Report then goes on to consider the 'relationships between proposals for mathematics and science'. The main aim of this is to satisfy the Secretary of State that the proposals for mathematics and science are in line with each other, and (see para. 3.23) the Report is satisfied in this respect. It next identifies a list of ATs in science 'which will contribute to or supplement those in mathematics'. This list of topics is related to measurement, in particular 'time' (see paras 3.24 and 3.25).

One might expect that the explicit promise of 'guidance on cross-curricular matters' would have been fulfilled to some extent on publication of the revised NC in mathematics (DES, 1991) and the accompanying non-statutory guidance (NCC, 1991), or at least readdressed. This is not the case. As indicated above, NCC (1991) refers the reader directly to the earlier non-statutory guidance (NCC, 1989) after indicating the basic changes between the two versions of the curriculum. Two documents published by NCC (1992a, 1992b), offering guidance to teachers concerning Ma1 and in-service help with this AT, once again offer no specific examples from other subject areas. The list of topics within the NCC Consultation Report (NCC, 1988) noted above cannot be seen as a genuine attempt to provide an example for cross-curricular work. It refers to one specific area of the mathematics curriculum, the list is drawn explicitly from the science curriculum (physics in particular), and regards the items as supplementing the mathematics curriculum, which is viewed as deficient in this area.

Not only is the non-statutory guidance (NCC, 1989) remiss in not providing exemplars of cross-curricular work, but it neglects to make use of opportunities which it creates for such exemplars. Thus in discussing two different teaching approaches or 'strategies' (NCC, 1989), the idea of fitting unit cubes together in a variety of ways and relating the volume of the shapes formed to their surface area is considered. The Statements of Attainment (SoA) within the mathematics curriculum which might be addressed when pupils take part in such an activity are analysed in full. However, completely absent from the analysis is any attempt to relate the results of the investigation to other areas of the curriculum. Thus, for example, small mammals such as mice and voles have a large surface area to volume ratio and hence must either eat a great deal in relation to their body weight, or be well insulated against losing heat by the use of fur, or both. On the other hand, large mammals such as elephants have the reverse problem. They need to lose heat and so have wrinkly skin and huge ears to increase the surface area through which body heat can be lost. Again, worms breathe through their skin, hence they need to maximize their body surface area, and their shape achieves that; the investigation referred to indicates that 'linear' constructions of cubes provide the maximum surface area. While the above information is easily available from biology texts, it is sadly absent in the non-statutory guidance.

From the complete lack of help for teachers and of exemplars in publications, one is forced to conclude that the integration of mathematics into the wider curriculum is not a serious intention of curriculum writers in mathematics. Mathematics is usually viewed as a tool in the crude utilitarian sense. Learning to use the tool takes place in an applications vacuum or through the use of 'problems of whimsy' (Pollak, 1969). All that would seem to matter is that 'In general pupils will meet and become secure with a particular mathematical idea or process at least one level earlier than they are required to use it in *science*' (NCC, 1988, my emphasis). The needs of other subjects have lip-service paid to them but no detailed consideration or exemplification given. The wider view of mathematics as a means of communicating, reasoning and representing through the idea of a mathematical model would seem to have been ignored or replaced by unhelpful rhetoric.

The second point which emerges is that the writers of the various reports do not seem clear about what is meant by the terms 'real' and 'real world'. No attempt is made to define either. The words frequently appear, as here, in quotation marks when first introduced into a section, only for these marks to be removed later on as though they had been defined by the intervening discussion. In fact no such process of definition ever takes place; for an example of this see NCC (1989, p. D4). There is clearly confusion as to what constitutes the real world, shown by the reluctance of the authors to define this concept or differentiate it in some way. By sheltering behind this undefined and non-differentiated concept, the authors of the various documents are able to issue platitudes and treat the problem of integrating mathematics into the wider curriculum as simplistic and non-problematic.

The above also indicates that there are two perceptions of the usefulness of mathematics that contribute towards making it an essential part of the curriculum in its own right and a growing component of a wide range of subjects across the curriculum. These are:

- mathematics as a tool;
- mathematics as a 'means of communication – to represent, explain and predict' (Cockcroft, 1982).

MATHEMATICS AS A TOOL

The perception of 'mathematics as a tool' is based upon an analogy, an analogy which is a safe and reassuring one. Everybody knows what it means because everybody uses tools of some description, in much the same way that everybody knows about teaching and what goes on, or should go on, in schools because everybody has been to school. A picture needs to be hung on a wall, so get a hammer, a couple of nails and the job is done; the shopping bill needs checking, so get out the 'old addition of columns of figures'. However, a moment's consideration of both situations shows such a simplistic comparison is fallacious.

In thinking about hanging the picture, would we not consider the condition of the wall? It might be possible simply to hammer in the nails; on the other hand, it might be more advisable to drill and plug the two holes and use screws. In our own houses we know which it is better to do and which it is possible to do for which walls, but in terms of the analogy, this information is taken for granted; 'reach for the hammer' is the message. Hence the interpretation of using a tool is somewhat crude and utilitarian, and so, through the analogy, is our interpretation of using mathematics.

To return to the other half of the example, the checking of the shopping bill, an example so beloved of official documents: is it sensible to add up the bill to check the total? Virtually all shops and stores give till receipts with the price of each item listed, the total due and the change required for the money tendered. Thus the total is guaranteed to be accurate in terms of the sums entered, so to check the addition of the bill is a waste of time. What needs to be checked is the separate amounts which have been entered into the till; that is, the initial data needs checking against the information available from the purchases themselves. However, the increasing use of bar-code readers and tills which operate on item prices means that the prices of individual articles are often not available for scrutiny once the shop has been left. Thus what is needed is not a tool but a form of thinking that is statistical and inferential in nature. Is this what is usually paid? Does it differ significantly from previous similar bills? If so, what item or items have changed in price? Thus addition is not required to check the bill, but may be required to anticipate, or estimate, the size of it. In a take-away food shop where the prices are usually displayed, one can mentally calculate the cost of the order and compare the total arrived at with that from the till. There are, however, many ways of estimating the total of such bills without resorting to cumbersome use of a mathematical tool, but rather using the same kind of thinking that is outlined above.

An example of this kind of thinking is the method that a friend uses to estimate the cost of a trolley-full of supermarket shopping. She counts the items as she places them in the trolley; the number of items is then the estimate for the cost in pounds of the shopping – seventy items are estimated to cost £70. The resulting estimate is, I have found in doing my own shopping, often remarkably close. As mathematicians we may wonder what my friend will do when inflation causes prices to rise. Perhaps we would propose that she revise her unit price upwards, say to £1.10. On the contrary, I would suggest that she will adopt a far less demanding strategy of maintaining her current practice but then adding on a fixed amount at the end. Thus seventy items would have an estimated cost of not $70 \times £1.10 = £77$, but $£70 + £10 = £80$. Given the level of accuracy being worked to against the mental effort required, this latter strategy seems far more likely to be used and to prove to be a sufficiently close approximation.

This is but one example of the failure of the analogy, and in one quite specific area. The analogy may well hold up in other circumstances. It is therefore necessary to consider it across the spectrum which mathematics education in schools has to address – generally consisting of three different but overlapping areas:

- 'preparing students for their private and social lives as individual citizens' (Niss, 1989), what might be loosely termed *the everyday world*;
- preparing students for their working lives, wherein they may or may not use that mathematics and what is used may change markedly at different times in their careers, namely *the world of work*;
- preparing students for their present and future academic lives, where they will meet and use mathematics in other subjects and be educated in mathematics itself, namely *the academic world*.

The everyday world

Sewell (1981) and the Advisory Council for Adult and Continuing Education (ACACE, 1982) have shown that many adults, including the well educated, are incapable of operating with the mathematics provided for them by their mathematical education in school. Indeed, the difficulty that Sewell had in persuading subjects to take part in the research shows the degree of fear concerning school mathematics and the degree of its rejection. Nevertheless, there is little doubt that such people do cope in their everyday life. Included in their strategies for coping may be those of avoidance, and where this is so there is no doubt that they cannot be fully participating members of their society. However, the work of Lave (1988) indicates that her subjects were successful in solving similar problems, for example which is the best buy for different sizes of product package, in the actual situation where the problem arose. This stands in marked contrast to the result from Sewell's subjects, who were generally unable to cope with a pencil-and-paper calculation concerning the better buy between two bottles of tomato ketchup during a more formal interview situation, which took place away from the real locus of the problem.

Context and situation are therefore vital to the way in which the problem is viewed and might be tackled However, the way that the problem is conceptualized is also changed as the problem is approached. As Lave (1988) puts it:

> This process of transformation of quantitative relations may be described as 'dissolving' problems (in both senses of the term), making them disappear into solution within the ongoing activity rather than 'being solved'.

What is crucial to many of the instances which Lave quotes from her own research and that of others is the way that people in everyday life use mathematical thinking to transform the problem into one which best reflects their own aims in finding a solution, and is economical in time and effort. This invariably seems to mean that the standard mathematical tool 'prescribed' by school mathematics is avoided, because either it is of no use or it is impractical to use or it can only be imperfectly remembered and understood.

An example drawn from Lave (1988) exemplifies this whole process. A new member of Weight Watchers was asked to prepare a serving of cottage cheese that was three-quarters of the standard allowance of two-thirds of a cup:

The problem solver in this example began the task muttering that he had taken a calculus course in college (an acknowledgement of the discrepancy between school math prescriptions and his present circumstances). Then after a pause he suddenly announced that he had 'got it!' From then on he appeared certain he was correct, even before carrying out the procedure. He filled a measuring cup two-thirds full of cottage cheese, dumped it out on a cutting board, patted it into a circle, marked a cross on it, scooped away one quadrant and served the rest. Thus, 'take three-quarters of two-thirds of a cup of cottage cheese' was not just the problem statement but also the solution to the problem and the procedure for solving it.

We are assured by Lave that at no time did the subject attempt to use the standard mathematical tool,

$$\frac{3}{4} \times \frac{2}{3} = \frac{1}{2} \, .$$

By thinking precisely about the meaning of the words in the problem, their mathematical interpretation in that context, the need for any mathematical tool was subverted. Indeed, it is doubtful if the subject would have been as certain of the correctness of his answer if he had used the formal method. The subject had an understanding, in context, of three-quarters of two-thirds; there are indications that he did not have an understanding of the formal algorithm or tool.

The problem of assigning meaning to symbolism is one which has been commented upon by Onslow (1991). The implication of the article, however, is that given the symbolism, it is important to be able to find a real-world context for it: 'When one cannot provide a representation for mathematical symbolism then, generally, one does not truly understand that which the symbolism portrays.' In other words the mathematics would seem to come first; it represents an abstraction from the real world and has a power which should be used. While not disputing the abstraction and the power, the compulsion is surely wrong. Indeed, as Onslow does acknowledge, 'Being a flexible problem-solver, who is able to adapt mathematical knowledge to a real world situation, saves both time and effort.' But this is precisely the point; the tools which mathematics offers do not save time and effort in the everyday life of most people. They are too abstract, too general; by fitting every situation they lack meaning within the context of the specific situation. Within the context of seeming to say to people, 'Here is the mathematical tool that you need for this situation', we appear amazingly arrogant. What we should be looking at is not the tools which we seem so keen to provide and our clients so keen to do without, but the mathematical thinking which goes on in attempting to describe the situation so that problem and solution become one.

The world of work

The mathematical requirements of the world of work have already been discussed in Chapter 2. The changing demands and the nature of the change described therein illustrate that within that particular world, the mathematical tools of the classroom are often inadequate or inappropriate. Rather it is suggested that problem-solving skills of the kind highlighted by Lave (1988), which are able to exploit the mathematical knowledge, the context and the experience of the worker, are to be more highly prized, together with the ability to think in an inferential and statistical manner.

However, as Chapter 2 emphasizes, demands change, and one of the results of that process is greater mathematical demands being made upon a smaller number of employees. In areas where the level of these demands is high, then they must be regarded as professional, and we should expect that formal training or education via the company, university, college or professional body will be provided. In this area the writings of Pollak (1969, 1978, 1988) are instructive, particularly the last reference. Pollak opened his contribution to the International Commission on Mathematical Instruction (ICMI) Symposium at Udine, Italy, of 1978 with a series of anecdotes, many of which concern his experiences as a research mathematician with the Bell Telephone Company. In many cases these experiences relate to engineers who would appear to regard mathematics as a tool. The pitfalls in taking such a stance are vividly illustrated by, for example, valid mathematics applied in the wrong circumstances, failure to understand the concept of proof and the role of counter-examples, and the inability to use available mathematics in order not to waste time with problems that are impossible to solve.

Combining the remarks of Pollak with those of Wain in Chapter 2 illustrates vividly that what is required by the world of work is the ability to think inferentially and statistically as well as mathematically, and to use these processes in one's career and professional development.

The academic world

The apparent separation of subjects by the NC into distinct areas of knowledge has already been noted above. However, as Wain (1983) suggests, such a curriculum model is easier to put into practice than an integrated curriculum. The latter, at a very basic level, requires an understanding of many fields of knowledge, their different sociologies and natures, and a willingness to be prepared to work against a background of uncertainty which may need continuous resolution. How difficult this is, especially in relation to mathematics, is shown by the discussion paper of Alexander *et al.* (1992) and the report of the NCC (1993) into primary education, where such a curriculum is attempted. The suggestion put forward to remedy this difficulty is that the classes of older pupils should be taught mathematics by specialist teachers, thereby emphasizing the separateness of mathematics and paving the way for the traditional divisions already well established within the secondary school.

Such divisions do much to emphasize the crude, utilitarian interpretation of mathematics, which disregards the need for interpretation and analysis of context within a given situation. Two personal experiences help to illustrate this.

In the first instance I was approached by a year-12 student to check the calculation of Spearman's Coefficient of Rank Correlation; as she put it, 'The answer should be about 0.9, but it isn't. It's 0.2. Have I done it wrong?' Since the calculation was part of a piece of geography course work, it is interesting in itself that I, a mathematician, should be approached to adjudicate upon the correctness of the use of the particular tool. The calculation was indeed correct. I next tried to establish the reason for doing the calculation and discovered that, in the context of the data which had been recorded, this was standard procedure. The final question then was whether or not the data had been correctly recorded. Again the answer was apparently in the affirmative. However, when the suggestion was made that the work be written up including the data and calculation, together with some reasons as to why the standard result had not materialized, which related to the location from which the data had been taken, the student put up strong resistance. From her point of view the data were wrong and the answer should have been 0.9 – it always was. A few weeks later, we met again in the corridor, and I asked if she had resolved her problem. She had indeed resolved it; she had been to see the head of geography, who had agreed that the data were wrong and had sent her away to gather some more. This she had duly done and the answer was now very nearly 0.9, sufficient to be judged correct by herself and the head of geography.

In this case the proper tool had been selected from the armoury of the pupil and correctly applied. The results, however, were not understood to be correct and so the data on which they were based were called into question. The idea that there was room for interpretation of the results within the context of the collection of the data was not to be entertained. The strict separation of the two subjects meant that interpretation of the results of a series of mathematical operations in terms of the context of those operations was not open to the student, though it clearly should have been.

In the second example, at another school where the mathematics staff taught a number of service courses for geography, business studies and biology, the head of geography requested that the mathematics for geography course be taught by the geography department. He had surveyed his students about the geography course as a whole and this was one of the unanimous responses towards improvement of the course from their point of view. The students could do the mathematics taught, mainly statistics, and indeed felt very confident about it. However, they did not have a clue as to when to use it in their own subject area. In consequence the geography department were having to teach it all over again. A word with the member of staff from the mathematics department who was teaching the course revealed that she was not using any examples from geography and was working from a standard mathematics textbook. Moreover, she knew no geography and no circumstances in geography when the techniques which she was teaching might be used.

In this case, students had a whole armoury of techniques or tools which they could use confidently within the confines of a mathematics textbook. Within their own subject area, they were unable to use any of the techniques, since they did not recognize the contexts in which they could be used. Separation of subjects had, in this case, caused the separation of tool and context of application. The confident use of the tools within the realms of the mathematics textbook was acceptable as evidence of the students acquiring the tools – until the students themselves were asked for their view about the matter.

One should not be too hasty to judge any of the teachers involved in these incidents. There are few indeed who would feel confident to teach both mathematics and another subject such as geography up to A-level. The separated curriculum has made specialists of us all, often to the detriment of our general education.

The conception of mathematics as a tool is therefore one which provides a crude, utilitarian view of mathematics, a view which does a disservice not only to mathematics itself, but to the student of mathematics as well. In none of the areas served by mathematical education is the student offered the kind of education in mathematics which is useful, relevant to the student and memorable. It is therefore of little wonder that so many people have such strong negative feelings about mathematics.

MATHEMATICS AS A MEANS OF COMMUNICATION – TO REPRESENT, EXPLAIN AND PREDICT

It has been argued above that to think of mathematics as a tool within and across the curriculum is a poor analogy which leads to a conception of mathematics as crudely utilitarian and as such does not reflect the intellectual processes involved in its use, even at a very basic level. Thus it has been argued that what is required is a conception of mathematics as a way of thinking which is interpretative, reflective and inferential in nature. Such a conception is much more in line with the Cockcroft view expressed in the heading of this section.

However, there must be something for the mathematics to communicate, something to represent, explain and predict. This would seem to imply that students of the discipline have a use for the mathematics which is relevant to them, a problem which holds a real interest for them. The origin of this need or problem is not important; it is the reality to the students which is important, and it is this reality which perhaps is best called the 'real world'. By taking this view of the term 'the real world', there is no need to make the traditional distinction between pure and applied mathematics. The world from which the problem comes, be it from the world of mathematics or from the world outside the classroom or lecture theatre, is as real as any world to the person involved with the problem.

Such a view of mathematics is very close to that of Peirce (1956): 'Mathematics is the science of possibilities.' This view has been developed further by Ormell (1972, 1975, 1977) into the idea that all mathematics is applicable, either within mathematics itself or outside the subject. Such a view is also consistent with the ever-increasing range of areas of application of mathematics as well as the growing number of topics within mathematics finding applications outside the subject. Indeed, to attempt to argue that any branch of mathematics would be useless in terms of application is fatuous in the extreme.

Pursuing this argument, what we come to see is that it is not the particular area of mathematics nor the particular application which is important, but the nature of the process by which the mathematics is applied to the situation or problem in hand. This process is commonly termed mathematical modelling.

MATHEMATICAL MODELLING

The idea of mathematical modelling is not new and neither are attempts to introduce it into the mainstream curriculum, particularly at post-16 level. An early attempt, the Sixth Form Mathematics Curriculum Project directed by Ormell, has already been noted above. While the formal syllabus and examination no longer exist, the approach and philosophy are still carried on through the Mathematics Applicable Group (MAG), based at the University of East Anglia, under the direction of Ormell.

The Mechanics in Action Project (MAP), originating at the University of Leeds and subsequently spreading to the Universities of Manchester and Sheffield, has been instrumental in promoting mathematical modelling and real problem solving. It began as a means of promoting the teaching of Newtonian mechanics through the use of apparatus, the Leeds Mechanics Kit, and the posing of problems from the worlds of funfairs, engineering and sport which could be modelled via the apparatus and mathematically using mechanics. In this way the project hoped to persuade teachers and pupils that mechanics was not a dull, impractical subject dependent upon seemingly meaningless fictions such as weightless pieces of string and frictionless pulleys, but rather had its foundations and its applications in the modern world in which we all live.

MAP has been very successful in putting over its message, to the extent that it has either written materials for or has advised all the A-level mathematics projects of the early 1990s in Britain on their mechanics modules and components. In certain instances the central theme of MAP, mathematical modelling, has been adopted in other areas of the syllabuses.

The starting point of the MAP philosophy is the modelling diagram (Figure 11.1) taken from Jagger *et al.* (1990). The diagram is simple, but must not be interpreted simplistically. While each stage or operation within the process might seem to be self-explanatory, there are hidden assumptions as we move between the various stages which need to be taken into account.

Discussions of the modelling process frequently proceed in the abstract without reference to any specific problem. While every problem is unique and to some degree specific, the difficulties of the modelling process are often hidden if an example is not followed through at the same time. The reader is therefore encouraged to consider the problem 'How fast can you walk?' and think how this might be modelled.

The problem has been stated very baldly and directly, which highlights the first major

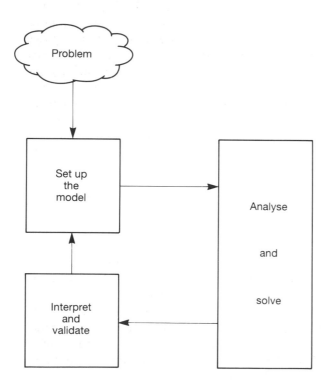

Figure 11.1   *MAP modelling diagram*

difficulty, that of moving from the real world, the locus of the problem, into the mathematical world of the model. The difficulty is often characterized as one of identifying the important variables. Potential variables are many: stride length, stride pattern, leg length, height of walker, weight and build of walker, ratios between pairs of these, and the nature of the surface being walked upon – smooth or rough, flat or sloping. The list is certainly long and only a few possibilities have been mentioned. However, there are more fundamental questions to be answered, which once settled upon often help to reduce the number and variety of potential variables. Among these questions are 'What kind of mathematics is to be used?' and 'What kind of answer is required?' The problem could be resolved by reference to statistics; an analysis of the performance of a large number of volunteer walkers or the past performances of race walkers could provide an answer. But is this the kind of answer being sought? Is an average, together with a standard deviation, sufficient? Perhaps a specific answer is required, for example $6 \, \text{km} \, \text{hr}^{-1}$, or a general answer, for example the speed v as a function of some determined variables. If the last of these is wanted, then the problem might be better approached through the medium of mechanics, which seeks to deal with the phenomena of motion through certain laws which can be expressed symbolically.

A third question, basic to the two above, is 'What is the meaning of the problem, in terms of both the individual words and the complete statement?' Therefore what does it mean 'to walk' and who is the 'you' of the problem – you the reader, the writer, or some generalized you?

All of these questions require answers or assumptions to be made if the problem is to be resolved in a satisfactory manner. These answers and assumptions serve to define the context of the problem, and, as seen above, the context serves to define the solution to the problem. In this case, solvers of the problem have the opportunity to set up the context for themselves rather than it being a given. For the novice, such choice is probably overwhelming, and because there may well be no grounds for making the choices one way or another, a position of power in fact becomes powerless. Wain (1983) points to the fact that knowledge of the area from which the problem comes can assist in making the best and most appropriate decisions, but that this knowledge may not be known by, and indeed may be counter to, the intuitions of the solver. Thus the appropriate model may not be obvious to the solver and may indeed be unattainable.

If one accepts these two points, then it is perhaps best if modelling is introduced within an area of mathematics where the same model can solve a wide variety of different, if related, problems and in which the basic assumptions are well known but capable of relaxation. Newtonian mechanics provides an ideal vehicle in all of these respects.

To return to the problem, if we choose to assume that the 'you' is a generalized you, that what is required is some form of general answer expressed symbolically, and that the technical meaning of 'to walk' is always having one foot in contact with the ground, then the area of mathematics which most suggests itself is mechanics. The choices to be made now are between particle and rigid body mechanics and between the numerous different kinds of motion which might apply. Again the novice is faced with a degree of choice that is overwhelming, or, if for instance only particle mechanics is known, with a leap of faith that the motion of the human body in such a complex motion as walking can be sensibly represented in some way by the motion of a particle.

Figure 11.2, adapted from Alexander (1992), shows the model that is proposed for

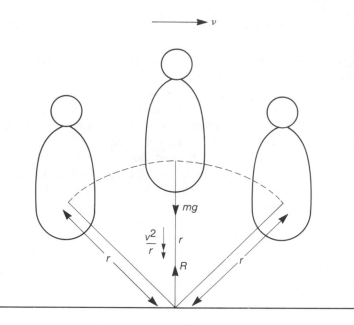

Figure 11.2   *A model for human motion*

the problem. If we imagine the knee joint to be locked, then the hip joint moves in an arc of a circle whose radius is equal to the length of the leg. The centre of gravity is just a few centimetres above hip joint and now also moves in circles of the same radius, r. The forces acting through the centre of mass are the weight, mg, and the normal reaction of the ground being walked over upon the foot, R. In the position when the leg is vertical, if the person is moving forward with speed v, then the acceleration will be v^2/r, directed towards the centre of the circle. Therefore, using Newton's Second Law, we have

$$mg - R = \frac{mv^2}{r}\,. \qquad (11.1)$$

The second stage of the modelling process, as described by Figure 11.1, has now been entered. The model has been set up and now the mathematical analysis is about to begin. But before this happens, the reader should reflect upon the model being used and how it has been derived. In its finished form it has an 'obvious' quality about it, but is it obvious? The model is due to Professor R. McNeill Alexander, Professor of Zoology in the University of Leeds and an acknowledged expert on the motion of animals of all kinds and the modelling of such motions via Newtonian mechanics. As an expert, he has both an intimate knowledge of the area from which the problem comes and prolonged experience of applying mathematics within this area. Such knowledge and experience are not available to the novice, a point which needs to be borne in mind when modelling is advocated as a part of the mathematics curriculum at any level.

Returning to the mathematics of the problem, we are now in the area of traditional mathematics, the manipulation of formulae. Thus from equation (11.1), we obtain,

$$R = mg - \frac{mv^2}{r}. \tag{11.2}$$

For walking to take place, the formal definition was that contact must exist at all times between the walker and the ground. Mathematically this condition is,

$$R > 0,$$

which when combined with equation (11.2), gives

$$gr > v^2. \tag{11.3}$$

Notice that this solution has not been obtained entirely through the use of mathematics, but has had to rely upon knowledge from the area of application, namely the formal definition of walking.

The modelling process now moves into the third stage, interpretation and validation of the model. This is the stage in which the conception of mathematics as a way of thinking which is interpretative, reflective and inferential comes to the fore. What does the bundle of symbols in statement (11.3) actually mean in terms of the original problem posed? Does it provide a sensible answer to the problem? Here again we need knowledge from the area of application.

The value of g is roughly $10 \, \mathrm{m \, s^{-2}}$, and the leg length of a typical adult is about $0.9 \, \mathrm{m}$. These figures give a value for v of $3 \, \mathrm{m \, s^{-1}}$, which is about the speed at which it becomes more economical in energy terms to run rather than to walk. We therefore have an answer which is sensible, but what does the formula of (11.3) tell us about walking? How can it be interpreted?

The maximum value of v is dependent upon two variables, the leg length and the value of g. If the leg length is reduced, then so is v. Consider walking with a young child. When you walk quickly, the child has to run to keep up. The length of its legs will not allow it to walk as fast as you. Similarly, recall the moving pictures of men as they walked upon the surface of the moon. In most cases they bounced along, rather than walked, and again the explanation is now clear. The value of g on the moon is less than on earth, hence the maximum speed is lower than on the earth; so much so that in trying to walk at normal speeds the astronauts bounced.

Having seen that the model is valid and that it is possible to use it to interpret other related facts, the next move in the modelling process is to return to the original model and problem and see if, in the light of the results from the model and the deeper understanding gained of the original problem, a second model can be produced, perhaps a refinement of the first. This meets a potential double barrier; the need for more knowledge of mathematics and of the area of application. To have modelled the situation successfully once does not guarantee that it can be refined or remodelled on the basis of the original information. Further refinements upon the particular problem of walking can be found in Alexander (1992) and Roper (1990).

There are therefore clear difficulties in the teaching and learning of mathematical modelling, difficulties which often do not seem to be admitted by enthusiasts of modelling. This once again illustrates the problem of integrating mathematics into the wider curriculum. In particular, there is a need for sufficient knowledge to be available about the area of application, and for students to have sufficient confidence in the mathematics known to use it in an unfamiliar situation.

However, MAP has also tried to introduce the ideas of mathematical modelling into the 11–16 mathematics curriculum, using mechanics as a source of material. The Universities of Manchester and Sheffield have been closely involved with these developments. The modelling used in them has been of a more numerical and experimental nature than the theoretical modelling described above. An excellent example of this form of modelling to solve a real problem, and one which also illustrates the use of statistics, is Taylor and Rouncefield (1989).

CONCLUSION

The integration of mathematics into the wider curriculum is not an easy task. It is difficult to foresee it happening in the near future, and yet there are already in place several of the prerequisites for it to come about.

The NC in mathematics has made the teaching of the process part of mathematics compulsory throughout the 5–16 age range through Ma1, 'Using and applying mathematics'. All children should therefore be involved, from a very early age, in using mathematics to solve real problems. Within the non-statutory guidance, a way of organizing the teaching and learning that depends to a large extent upon the sympathetic use of problem solving is advocated. This method makes the need to solve problems the motivation for the study of the mathematics involved.

Groups such as MAP have attempted to adopt similar stances and have gone much further than the NCC in providing classroom materials, support and in-service training for teachers who wish to work with their materials. Much work has been done with local groups in the Manchester and Sheffield areas supported by funds from LEAs, TVEI and charitable trusts.

Other local initiatives have included curriculum audits whereby individual schools have investigated where and when in the curriculum particular topics are taught, the aim being to see whether the teaching of the whole of the NC can be rationalized in some respects. Schools also hold project or theme weeks during which a whole year group of students is involved in looking at a common theme in all subject areas. Thus the topic of bridges might be addressed across the whole of the school's curriculum for 14-year-old pupils.

What may well prevent many of these initiatives from taking root and spreading into true integration of the whole curriculum, not just mathematics, is the NC and its attendant assessment arrangements. That which is to be assessed formally, namely ATs 2–5, and indeed only those SoA therein which are assessable by timed written examinations, may well be what is taught and given emphasis in the classroom. HMI (1992) note the increase in restrictive educational practices in the classroom and in testing.

At a time when technology is reducing the need for manipulative techniques and making the analogy that mathematics is a tool in the utilitarian sense even more difficult to sustain, we appear to be becoming locked into a curriculum which will value manipulative techniques and the tool mentality simply because this is what is to be tested. The interpretative, reflective and inferential view of mathematics appears to be one which, although vital to mathematics and to its use in other subjects, and the only view sustainable in the face of new technology, is to be neglected in the 5–16 age range.

It is therefore fortunate indeed that advances within A-level mathematics syllabuses, outlined above, may well force the education of mathematicians into a much broader curriculum. This is being recognized at university level, where undergraduate courses are beginning to appreciate that there is a need to introduce modelling and course work, that the starting point of many of the undergraduates is different from what it was in the past, and that undergraduates now bring different skills with them to their courses. Many universities are moving into modular structures which will also allow the students to broaden their education. Perhaps the integration of mathematics into the wider curriculum will have to wait until there are teachers available whose experiences and education are sufficiently different from what they see before them in the schools for them to feel compelled to take the necessary steps themselves.

REFERENCES

ACACE (1982) *Adults' Mathematical Ability and Performance*. Leicester: ACACE.

Alexander, R. McNeill (1992) *The Human Machine*. London: Natural History Museum Publications.

Alexander, R., Rose, J. and Woodhead, C. (1992) *Curriculum Organisation and Classroom Practice in Primary Schools. A Discussion Paper*. London: DES.

Cockcroft, W. H. (1982) *Mathematics Counts*. London: HMSO.

DES (1988) *Mathematics for ages 5 to 16*. London: HMSO.

DES (1989) *Mathematics in the National Curriculum*. London: HMSO.

DES (1991) *Mathematics in the National Curriculum (1991)*. London: HMSO.

Fitzgerald, A. (1985) *New Technology and Mathematics in Employment. The Main Report*. Birmingham: University of Birmingham.

Fitzgerald, A., Purdy, D. W. and Rich, K. M. (1981) *Mathematics in Employment (16–18), Report*. Bath: University of Bath.

HMI (1985) *Mathematics from 5 to 16*. London: HMSO.

HMI (1992) *The Implementation of the Requirements of the Education Reform Act. Mathematics. Key stages 1, 2 and 3. A Report by H.M. Inspectorate on the Second Year 1990–91*. London: HMSO.

Holt, M. and Marjoram, D. T. E. (1973) *Mathematics in a Changing World*. London: Heinemann.

Howson, A. G. and McClone, R. R. (1983) *Mathematics at Work*. London: Heinemann.

Jagger, J. M., Roper, T. and Savage, M. D. (1990) *The Teacher's Guide to the Leeds Mechanics Kit*. Leeds: The Mechanics in Action Project.

Lave, J. (1988) *Cognition in Practice*. Cambridge: Cambridge University Press.

Mathematics Applicable (1975) *Mathematics Changes Gear. Students' Discussion Unit*. London: Heinemann Educational Books/Schools Council.

NCC (1988) *National Curriculum Council Consultation Report. Mathematics*. York: NCC.

NCC (1989) *Mathematics Non-statutory Guidance*. York: NCC.

NCC (1991) *Mathematics Non-statutory Guidance (1991)*. York: NCC.

NCC (1992a) *Using and Applying Mathematics Book A. Notes for Teachers at Key Stages 1–4*. York: NCC.

NCC (1992b) *Using and Applying Mathematics Book B. INSET Handbook for Key Stages 1–4*. York: NCC.

NCC (1993) *The National Curriculum at Key Stages 1 and 2. Advice to the Secretary of State for Education*. York: NCC.

Niss, M. (1989) Aims and scope of applications and modelling in mathematics curricula. In W. Blum, J. S. Berry, R. Biehler, I. D. Huntley, G. Kaiser-Messmer and L. Profke (eds), *Applications and Modelling in Learning and Teaching Mathematics*. Chichester: Ellis Horwood.

Noss, R. (1990) The National Curriculum and mathematics: a case of divide and rule? In P. C. Dowling and R. Noss (eds), *Mathematics versus the National Curriculum*. London: Falmer Press.

Onslow, B. (1991) Linking reality and symbolism: a primary function of mathematics education. *For the Learning of Mathematics* 11, 33-7.

Ormell, C. P. (1972) Mathematics, applicable versus pure-and-applied. *International Journal of Mathematics Education in Science and Technology* 3, 125-31.

Ormell, C. P. (1975) Towards a naturalistic mathematics in the sixth form. *Physics Education* 10, 349-53.

Ormell, C. P. (1977) A new naturalistic style of mathematics. *Physics Education* 12, 257-60.

Peirce, C. S. (1956) The essence of mathematics. In J. R. Newman (ed.), *The World of Mathematics: 1773-1783*. London: Allen and Unwin.

Pollak, H. O. (1969) How can we teach applications of mathematics? *Educational Studies in Mathematics* 2, 393-404.

Pollak, H. O. (1978) The interaction between mathematics and other school subjects. In International Commission on Mathematical Instruction (ICMI), *New Trends in Mathematics Teaching. Vol. IV*. Paris: UNESCO.

Pollak, H. O. (1988) Mathematics as a service subject – why? In A. G. Howson, J.-P. Kahane, P. Lauginie and E. de Turcheim (eds), *Mathematics as a Service Subject*. Cambridge: Cambridge University Press.

Roper, T. (1990) Mathematics and motion of the human body. *The Mathematical Gazette* 74, 19-26.

Sewell, B. (1981) *Use of Mathematics by Adults in Daily Life*. London: ACACE.

Sutton, O. G. (1954) *Mathematics in Action*. London: Bell and Hyman.

Taylor, P. and Rouncefield, M. (1989) Swings, strings and statistics. *Mathematics in School* 18(4), 13-15.

Wain, G. T. (1983) Mathematics and other subjects. *Educational Analysis* 5, 45-55.

Chapter 12

New Technology and Mathematics Education: New Secondary Directions

John Monaghan

INTRODUCTION

Computer-based technology is changing the character of mathematics. Side by side with the rise of computers has been the increasing importance of discrete mathematics. This is examined separately in Chapter 5. Computers not only introduce new areas of mathematics but bring with them new ways of thinking about mathematics. Recursion is a good example. It is certainly not new to mathematics, but the emphasis on recursion within mathematics is new in areas such as logic, numerical analysis and the solution of equations. Fractals and the new science of chaos (see Chapter 6) are built on recursive procedures. They are also a crucial feature of Logo – showing that mathematics teaching is now inculcating recursive procedures as a fundamental mathematical tool for young students.

The ability of computers to crunch very large numbers accurately has opened up new areas within statistics and the solution of differential equations. Differential equations, moreover, can now be analysed graphically – a route that was hitherto impractical. Indeed, computer graphics open up areas for mathematical analysis far beyond that of functions in a Cartesian frame. The whole area of proof within mathematics is changing too. Computers can generate proofs of theorems. Often these proofs utilize a large number of cases, and *algorithmic* proofs are emerging. For further discussion of all of these changes, see Atiyah (1984), Cornu (1988) and Dubinsky and Tall (1991).

In England and Wales, aspects of this revolution in mathematics are evident in the National Curriculum (NC: see DES, 1991). The curriculum for the age range 5–16 now includes Logo, basic programming, spreadsheets, databases, graphic packages and statistics packages. The changes should not be exaggerated, however, as most of the computer-based suggestions are optional within the NC. In contrast, the changes highlighted above are beginning to filter down to upper secondary mathematics. It is this area that is examined in this chapter, and in particular the use of computer graphic systems (CGSs), supercalculators and computer algebra systems (CASs).

The technical scene is surveyed briefly, as this is liable to rapid change. Curriculum implications are then considered. Finally, and most importantly, evidence for how students learn in these environments is examined.

THE TECHNICAL SCENE

Computer graphics are not new. Most old microcomputers supported high resolution graphics, and for over ten years (at the time of writing) school teachers have used these facilities to write graph-drawing programs. In England some A-level schemes (Mathematics in Education and Industry (MEI), 1983; School Mathematics Project (SMP) – see Corlett, 1985) brought out suites of software to support their courses. These software suites included graphical demonstrations of selected A-level topics. General purpose mathematical graph-drawing packages came on the market at about the same time. The most popular of these were *Function Graph Plotter* (Association of Teachers of Mathematics (ATM), 1982) and *Supergraph* (Tall, 1985). These programs allowed students to explore a wide range of mathematics for themselves without the restrictions forced by a suite of programs tied to demonstrating a particular process. They were generic in the sense that they could be used for many purposes. They could also be used for pre-A-level mathematics. Tall extended his software in *Graphic Calculus I, II & III* (Tall, 1986a), which used the graphics features of *Supergraph* to create a graphics environment for the differential calculus, the integral calculus and differential equations.

Software is the key to extending the computer revolution in mathematics but new software development depends on suitable hardware for support. The machines developed in the late 1970s and early 1980s, although still evident in large numbers in British schools, are not the machines of the future. PCs, Apple Macs and, in Britain, the Archimedes, with fast processors, generic software support, a mouse and some form of Windows environment, are the immediate future. *Omnigraph* (Brayne, 1988) is an example of a graphics system for these new machines. It has built-in features to transform graphs, draw tangents and normals, draw the gradient and area functions, display vector fields of differential equations and generate power series. As will be seen below, however, there are other software products with similar graphic capabilities that incorporate symbolic manipulation facilities as well. Before these are examined it is useful to consider advances in hand-held technology.

Supercalculators is a term used by some in recent years to describe calculators with graphic and programming facilities. Programmable calculators have been around since the mid-1970s. Programming these was often a very cumbersome process. In the mid-1980s Casio, Hewlett-Packard, Sharp and Texas Instruments introduced easier programming and the ability for the calculator to display graphs. In England these were allowed in virtually all national mathematics examinations by 1990, which has certainly affected their importance in school mathematics, and they have now become a feature in new upper secondary mathematics courses such as SMP 16–19 and the Nuffield A-level Mathematics Project.

By far the most popular supercalculator in use in schools in Britain at the moment is the Casio fx-7000G, with variants. This is a cheap and moderately easy-to-use machine. It does not allow for any symbolic manipulation. Its popularity has allowed booklets to be written for its use, e.g. Reid (1990), and for curriculum experiments to be based on it, e.g. Ruthven (1990). In the early 1990s Casio brought out the fx-7700G with enhanced facilities, but this has not, to date, made the impact that its precursor did. In the early 1990s in Britain the Texas Instruments range (TI 81, TI 82 and TI 85) is starting to become popular. Apart from programming and graphics these machines offer more data types (matrices, vectors, lists, complex numbers), some symbolic manipulation (equation solving) and calculus-related graphic features.

Hewlett-Packard have offered the most impressive machines to date, but there are drawbacks which are worthy of comment. The HP 28C and HP 48SX offer some symbolic manipulation as well as graphics and programming, but the price is well over three times that of the Casio machines. If supercalculators are going to have a major impact on secondary education, then students will have to have their own machines and prices must be as low as possible. The second drawback of these Hewlett-Packard machines is that they are not easy to use. Different operations require different modes and the stack controls the order of operations. The procedure required to obtain a result can be very difficult.

Recently palmtop PCs with built-in word-processor, calculator, equation-solver, notebook and spreadsheet have appeared, and the immediate future offers handheld pen computers. These small machines allow ROM cards to be inserted so that more advanced mathematical tools, such as *Derive* (see below), can be added. This may well be the long-term future but at the moment the price puts them well beyond the budget of most students and schools.

Computer algebra systems or symbol manipulators

Computers have been able to manipulate lists of symbols for about thirty years (Howson and Kahane, 1986), but symbol manipulators really arrived about twenty years ago. There are now a number of systems of which REDUCE, MACSYMA, *Derive, Maple* and *Mathematica* are the main ones in use. A description and comparison of the facilities of these systems, up to mid-1991, is contained in Harper *et al.* (1991). REDUCE now has many users and an extensive library of procedures, but it is really only for university level, as the user needs a mature mathematical understanding in order to define functions and procedures to supplement the basic built-in functions. MACSYMA, too, is more suited for university than for school use and is viewed largely as a tool for research rather than for teaching. It also has some oddities for English-language users, e.g. the French '*entier*' occurs in place of 'int' for 'integer part'. The other three systems are all potentially suitable for use in schools. It is worth examining some of their features.

Maple will run on 386 PCs, Apple Macs and a range of computers not found in most schools. It requires at least 2 MB, preferably 4 MB, of memory and 7 MB of disk space. *Mathematica* runs on a similar range of machines and needs 5 MB of memory, 8 MB preferred, and 7 MB of disk space. *Derive* runs on PCs or computers that can emulate PCs (in Britain, for example, the Archimedes or Nimbus). It requires 512 KB of memory. At the time of writing *Derive* is by far the least expensive system and, if networks are to be used, the only one that a school could contemplate buying. However, both *Maple* and *Mathematica* have subsystems available as student versions for a similar price to *Derive*.

Derive is based on a menu/sub-menus system for everything including entering expressions. This makes it much easier for the beginner to use than either *Maple* or *Mathematica*, which require the user to input the correct syntax. Graphics on these two systems are, at the time of writing, easier to manipulate than on *Derive*, where there is no cursor control and no auto-scaling.

To illustrate the operations of each of these three systems, the table shows the coding and the output involved in expanding $(1 + x)^3$, differentiating the result and then factorizing the derivative. The menu options in *Derive* are indicated by square brackets. Thus [A]uthor indicates that menu option A (Author) is chosen.

	Derive	*Maple*	*Mathematica*
Input	[A]uthor $(1 + x) \wedge 3$ [E]xpand [C]alculus [D]ifferentiate [S]implify [F]actorize	$y: = (1 + x) \wedge 3;$ $y: = \text{expand}(y);$ $dy: = \text{diff}(y, x);$ $\text{factor}(dy);$	Expand$[(1 + x) \wedge 3]$ Differentiate$[\%, x]$ Factor$[\%]$
Output	$(1 + x)^3$ $x^3 + 3x^2 + 3x + 1$ $\dfrac{d}{dx}(x^3 + 3x^2 + 3x + 1)$ $3x^2 + 6x + 3$ $3(x + 1)^2$	$y: = (1 + x)^3$ $y: = 1 + 3x + 3x^2 + x^3$ $dy: = 3 + 6x + 3x^2$ $3(1 + x)^2$	In[1]: = Expand$[(1 + x) \wedge 3]$ Out[1] = $1 + 3x + 3x^2 + x^3$ In[2]: = Differentiate$[\%, x]$ Out[2] = $3 + 6x + 3x^2$ In[3]: = Factor$[\%]$ Out[3] = $3(1 + x)^2$

This technical scene is certain to change in the near future with new systems arriving and prices falling. It can only be hoped that designers will listen to those who use them in detailing future specifications.

A word about terminology is appropriate here. At present the terms computer algebra system (CAS) and symbol manipulator (SM) are used interchangeably. If a SM is viewed purely as a symbolic device, then a CAS is a SM with an accompanying computer graphics system (CGS). To avoid ambiguity these terms are so used in this chapter.

The rapidly changing technical scene creates problems for textbook writers, publishers and, ultimately, school teachers. Lyons (1988) discusses problems that publishers face in producing textbooks. Publishers produce for a market and cannot afford to make a loss. He argues that a market must be made for CASs before publishing houses, other than specialist ones, produce the textbooks teachers will require in a CAS-orientated future. In Britain, as mentioned above, the SMP 16–19 course makes essential use of graphic calculators. It gets over the problem of several different makes by including, as a supplement to its textbooks, a ring folder of alternative worksheets for different makes of calculator, which can be upgraded as new features appear. This is one solution. Another is to make reference to systems generic; that is, refer to a general feature that any system will have rather than particular keys or notation that a specific system will have. This clearly applies to supercalculators, CASs and even spreadsheets and other mathematical software.

Looking at the practicalities of textbook producers shows that change in practice is inextricably linked with the curriculum, which will now be considered.

SOME CURRICULUM ISSUES

Chapter 11 discusses the mathematics curriculum in global terms. This section examines local curriculum issues related to the technological changes described above. Due to the nature of the technology, curriculum issues are restricted to those to do with middle- and top-ability 14–19-year-olds. There must first be agreement as to what counts as the mathematics curriculum for this age and ability group of students. In England and Wales this can be argued to be Levels 5–10 in the mathematics NC document (DES, 1991) together with the core A-level syllabus of the Standing Conference on University Entrance and the Council for National Academic Awards (SCUE and CNAA, 1978). Within this there is a great deal of topic work open to a CGS or CAS approach; for instance, all the graphic and algebra topics required for, and including, fairly advanced calculus of a single variable. Rather than go through a list of topics, the discussion below focuses on two central areas and considers the effects that this technology has (could have) on this content before going on to discuss process issues in the curriculum.

The solution of equations

Any equation that can be expressed in the form $y = f(x)$ can be plotted by any CGS or supercalculator. In addition, most systems allow parametric and polar equations to be plotted. By plotting and zooming in on x-axis intercepts, very accurate numeric solutions of equations can be obtained. This allows curriculum planners to put less stress on techniques needed to solve specific types of equations and more on general features of the solution(s) to an equation, e.g. what it means for an equation to have or not have a solution, what this means graphically, and whether the solution is exact or approximate. This also allows more time for the application of the solution of equations to solving problems. Because the equations do not need to be easy to manipulate by hand, the problems can be real ones and not ones designed for specific algorithmic solutions.

Some supercalculators and all CASs can provide the solution of consistent sets of linear equations. This can be as a set of equations inputted as a vector, or by casting the set of equations in matrix form and using the built-in matrix operators to provide the solution in the same way as would be done with pencil and paper. In this way the solution of large sets of equations can be performed by students who have learned the method for solving pairs of equations. This has the advantage of increasing the range of applications of simultaneous linear equations and, by virtue of the speed of the machines, providing students with the time to experiment with a large number of sets of equations. At present the capabilities of supercalculators and *Derive* do not extend to sets of non-linear equations or inequalities.

Exact solutions to many equations can also be found by some supercalculators and all CASs. For supercalculators these are, at present, limited to polynomials. Given this fact, what emphasis should be placed on, say, the solution of quadratic equations by the formula or by factorizing? Quadratic equations are important for problems in kinematics, but physicists are not essentially concerned about the mathematical method of solving them and are more likely than mathematicians to be content with a machine that will give an accurate answer. In mathematics, quadratic equations are useful in that they are the first non-linear polynomial equations that students meet. They are also the easiest to solve – a fact that disappears given the new technology under discussion here.

Students have the power to consider a wide range of polynomials. Indeed, the need to concentrate on polynomials also disappears, since with CASs solving $2^x - 5 = 0$ is no more difficult than solving $x^2 - 5 = 0$. This is not an argument for allowing button pushing to replace thinking in the solution of equations, but an argument in favour of the mathematical community at large reconsidering aspects of the curriculum in the light of what can be done with this technology.

The means of solving these equations on present systems raises some further issues. Although a CAS can be instructed to solve $\sin x - x = 0$ or $x^5 = 1$, the user must often specify the precision – approximate or exact – and, when factorizing, the domain – rational or complex numbers. These are important qualitative aspects of the study of the solution of equations, often not dwelt on in courses that concentrate on the mechanics of obtaining solutions. New technology not only motivates these questions but also has the power to enable teachers to spend more time on them.

In England and Wales, calculators have already had an impact on this area of the curriculum. Iterative methods for the solution of equations, including the most basic, trial and improvement, are part of the NC (DES, 1991). Graphic calculators have, moreover, turned curve-sketching on its head – no longer do examination questions ask a student to sketch the curve of a function; the natural way to pose such questions now is to display a curve and ask the student to suggest a suitable algebraic formula for it.

Calculus

Graphic displays on any system clearly mean that much of what used to be taught as a mainly algebraic subject, because of the tedium of producing graphic results by hand, can be given a full graphic treatment. This is important as differentiation, for example, is a process that operates on a function, not simply an expression. It is possible to display a function graphically, plot values of the gradient and obtain its derivative as another graph, thus showing the qualitative meaning of differentiation. The power of graphics to display graphs, and to zoom in on points to find local maxima and minima, may also cause us to question the need for students to spend many hours studying algebraic techniques before they can tackle problems such as 'Find the maximum area of a right angled triangle with unit hypotenuse'.

Graphic systems can introduce new areas to the curriculum. A differentiable function is, when suitably magnified about a point, locally straight (Tall, 1986b, 1990). This enables us to speak of 'the gradient of a point on the curve' rather than 'the gradient of the tangent to a point on the curve', a subtle but an important distinction. This approach has been adopted by the SMP 16–19 course in England at present (SMP 16–19, 1991).

Graphic systems also open up new areas in the solution of differential equations. At present the 16–19 curriculum in England includes the solution of first and second order ordinary differential equations by separating variables and by integrating factor. Euler's approximation is also used to obtain approximate solutions. Using graphic systems allows students to dwell on the qualitative side of what the solution means and, indeed, whether there is a solution, as well as to examine solutions to equations that cannot be solved in elementary terms, e.g. $dy/dx = y^2 - x$ (see Tall and West, 1986). CASs can give analytic solutions to many types of ordinary differential equations but, at present, this area is less satisfactory than many other areas. In *Derive*, for example, the user cannot use the menu system to solve differential equations (except, of course, those that can be solved by direct integration). The *Derive* user must program the method of solution required or call on existing programs in the utility files provided. Considering, however, how much time is spent labouring techniques when many students do not even realize that the solution of a differential equation is a family of functions, this is an area ripe for curriculum change.

CASs have the power to affect radically the way in which students learn the basic ideas behind the calculus, for they allow not only symbolic differentiation and integration but also the evaluation of limits and sums. For example, to perform

$$\int_0^z x^2 \, dx$$

in *Derive* it is possible simply to key in the expression and integrate it:

Input	*Output*
[A]uthor $x \wedge 2$	1: x^2
[C]alculus [I]ntegrate	2: $\int_0^z x^2 \, dx$
0, z [S]implify	3: $\dfrac{z^3}{3}$

Etchells (1993a), however, shows how the power of a CAS can be used to enable students to integrate from first principles in a traditional infinite sum method:

Input	*Output*
[A]uthor $(rz/n) \wedge 2z/n$	1: $\dfrac{\left[\dfrac{rz}{n}\right]^2 z}{n}$
[C]alculus [S]um r	2: $\displaystyle\sum_{r=1}^{n} \dfrac{\left[\dfrac{rz}{n}\right]^2 z}{n}$
1, n [S]implify	3: $\dfrac{z^3 (n+1)(2n+1)}{6n^2}$
[C]alculus [L]imit n inf	4: $\displaystyle\lim_{n \to \infty} \dfrac{z^3 (n+1)(2n+1)}{6n^2}$
[S]implify	5: $\dfrac{z^3}{3}$

Thus this software allows a number of approaches to this topic and need not be simple button pushing.

Sequencing of topics

CGSs and CASs have the potential to change the way topics are delivered, the sequencing of topics and the weightings of sections within topic areas. This can be illustrated with the examples examined above.

An approach to quadratic equations with a CAS is to let the computer expand and factorize expressions. Students can then discover, and later prove, the rules for themselves and use the computer to check their predictions. Such an approach is described in Hunter and Monaghan (1993). Graphics windows can be used to show the connection between the factors of an expression and the roots of the corresponding equation. In all of this there is most definitely a place for the teacher, both in introducing early examples with simple factors/roots and in leading discussions with the groups or the whole class on what the students have found. In Hunter and Monaghan (1993) it was found that the usual 'teacher exposition–student exercises' approach was inverted, so that students discovered the rules for themselves and discussed them, with the teacher simply noting down the rules that they had obtained. This topic also allows a natural introduction to complex numbers as the roots to certain equations, just as basic calculators introduce children to negative numbers when they key in an expression like $3 - 7$.

Prior to CASs, the 'maximize the area of a right-angled triangle with unit hypotenuse' problem would be solved by setting up the equation $A = (x/2)\sqrt{(1 - x^2)}$, differentiating, equating the result to zero, solving the new equation and checking that it is a maximum. To differentiate this requires using both the chain and the product rule, both of which require many hours of tuition to master. With a CAS the differentiation can be performed by the computer. The students can still utilize the ideas of calculus but with the calculations performed by the computer. In this way they can quickly appreciate the power of the techniques.

Limits are undoubtedly one of the most difficult areas of mathematics (Cornu, 1991). The traditional approach to calculus is to teach limits before differentiation or integration. An alternative approach, using computers and the idea of local straightness, is to stagger the teaching of limits, since the graphic calculus ideas involved can be taught without students having a precise limit definition – indeed, the process can aid the development of limit concepts.

Algorithms

CASs cause us to rethink the place of traditional algorithms. What is the point of learning the formula for the solution of quadratic equations, how to factorize polynomials, simplifying trigonometric identities, the various differentiation and integration rules and a host of other routine manipulations, from simplifying algebraic fractions to simplifying transcendental expressions? To ask this question is not to question the need for students to learn these topics at all, but simply to ask how much they need to know and at what point they need to know it.

Shumway (1988) distinguishes between 'skill learning' and 'concept learning'. A *skill* is a procedure that is learned 'quickly, accurately and with little mental effort'. A *concept* is an abstraction 'that allows one to treat a class of objects or ideas in a similar manner for a specific purpose'. Much of what goes on in higher secondary algebra is learning skills. Are these needed when supercalculators and CASs can do the work for us? Even more to the point, will the relief from concentrating on routine skills free students and teachers to study concepts in more depth?

Certainly the point of doing any algorithm in mathematics is to solve problems. These problems may be pure – applicable to understanding mathematics itself – or applied – applicable to a modelling task. Algorithms may be applied to perform mathematical tasks in the best manner possible given the technology of any period of time; thus forty years ago square roots were extracted by hand. Teachers wish to place efficient algorithms in the hands/minds of their students, but in the process the algorithms often become ends in themselves. As a result the emphasis is placed on learning skills, not concepts. The reader need only reflect on typical lessons in any topic from factorizing quadratic expressions to integration – the vast majority of the time is spent doing routine examples.

A situation where students were using CASs to simplify $(2x + 4xy)/2x$ would be to the detriment of mathematics, this being analogous to younger students using a calculator to simplify $12 \div 3$ (except, perhaps, with very weak students who may need such a prop). But certainly students should be led to gain what Skemp (1977) called a relational understanding of why the algorithms work, as opposed to a purely instrumental understanding of the mechanics of applying algorithms. Bibby (1991) puts forward ways in which CASs can lead students from an instrumental to a relational understanding of algorithms. He argues that calling on new technology to perform an operation (instrumental understanding) can lead the student to *unpack* the algorithm, i.e. to discover how it works, which leads to relational understanding, This, however, is an area that needs more research.

It is quite possible that many of the algorithms taught at the moment are important for a relational understanding of an area of mathematics but are presently employed to rehearse instrumental skills. This argument is examined by Etchells (1993b) in the particular case of the chain rule for differentiation with respect to CASs. He argues that not only is the use of the chain rule as a skill made redundant by CASs, because the system will deliver the correct answer for any well-behaved function, but that students can also use a CAS to differentiate any such function from first principles. He goes on, however, to argue that there is an important place for the chain rule in a student's general appreciation of mathematics, because 'it is a one dimensional form of the general chain rule and it is important for solving differential equations via the

$$\frac{\mathrm{d}v}{\mathrm{d}t} = \left(\frac{\mathrm{d}v}{\mathrm{d}x}\right)\left(\frac{\mathrm{d}x}{\mathrm{d}t}\right)$$

relationship . . . the time that is not spent on drill and practice of the chain rule can be spent on concept development and justification of its validity'.

The calculator debate revisited

There are many similarities between the arguments voiced above and the arguments of calculator proponents in the calculator debate that has raged for several decades. As supercalculators are merely an extension of advanced scientific calculators, it is not surprising that many of the arguments for and against calculators translate to arguments for and against supercalculators. This translation extends to CASs.

This chapter is being written in 1993. Thirty years ago, mechanical calculators were being used by some mathematics teachers. Twenty years ago, the Schools Council (1972) devoted just one page to electronic calculators, as opposed to many pages on other calculating aids, in a book on arithmetic. Ten years ago the Cockcroft Report (Cockcroft, 1982) recommended that 'calculators should replace logarithm tables as the everyday aid to calculation'. Today, calculators are an everyday feature of mathematics lessons and logarithm tables are a feature of the past. Today, graphic calculators are being taken seriously but CASs are only advocated by a few and barely known by many. In ten years' time, who knows what the situation will be? Technology is leading education once again. It is useful to look at some of the changes brought about by the calculator revolution, and to project these to more advanced systems.

Calculators have made many processes redundant – extracting square roots, logarithm tables and, to some extent, long division. Clearly supercalculators and CASs have the same potential with regard to plotting graphs, simplifying algebraic expressions and differentiating complicated functions.

There is evidence that calculators are an aid in the process of solving problems (Bell *et al.*, 1983). The main reason put forward is that freeing students from time- and mind-consuming calculations allows them to focus on the type of mathematics the solution requires, i.e. more time determining whether a problem is an add or a times and less time on evaluating 2.34×3.45. This argument would appear to translate to supercalculators and could hold for CASs, though there is a real need for more research evidence before this could be stated with any certainty.

Calculators cause a resequencing of topics. Negative numbers, for instance, arise naturally when certain keys are pressed ($3 - 7$ instead of $7 - 3$), and decimal numbers similarly (when $3 \div 6$ is pressed instead of $6 \div 3$). This may not be at the time that a non-calculator curriculum would have introduced them. There is evidence, however (Shuard *et al.*, 1991), that these impromptu introductions can be important motivators for children to understand new areas of mathematics. In the same way π, i or e may arise in the CAS solution of an equation or, on a CGS or supercalculator, asymptotes may arise when 'tan' is keyed instead of 'sin'.

Calculators cause some topic areas to gain increased importance, e.g. iteration. Supercalulators have created a greater emphasis on graphical exploration of functions. CASs could well lead to increased use of matrix methods (all matrix operations are quite simple on a CAS).

It has been argued (Costello, 1992) that the calculator revolution is a failed revolution, in that mathematics educators have not responded to the potential to be freed from computational algorithms, and calculators have been used to support the teaching of the techniques they were to supplant. There are parallels here with the HP 48SX supercalculator. When this calculator differentiates, say, $y = x\sin x$, it produces

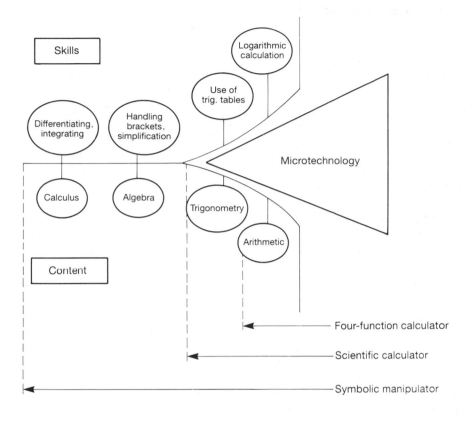

Figure 12.1 *The impact of microtechnology on mathematics teaching*

d/dx(x)sinx + x(d/dx(sinx)) before producing the final answer; that is, it repeats the very algorithm it replaces. The argument thus returns to the debate on algorithms. This debate seems certain to go on for many years for mathematics teaching is, in Bibby's (1991) terminology, just past the thin end of a microtechnology wedge (see Figure 12.1).

SUPERCALCULATORS, CASs AND STUDENT LEARNING

This chapter has, so far, outlined a number of things CGSs, supercalculators and CASs can do and possible ways students can use them. Now a mathematican picking up one of these systems immediately sees the potential. However, there are likely to be differences between how a mathematician uses these systems and how a student does simply because the mathematician knows where he or she is going. It is, therefore, important that as much research as possible is done on how students use and learn with these systems, for otherwise curricula may be set up that are fraught with difficulties and the revolution which seems imminent may fail. The first three research studies reported on below represent important studies done on CGSs, supercalculators and CASs. Further exploratory work on CASs is then briefly considered.

Computer graphic systems

As mentioned above, Tall developed a graphic approach to calculus using specially written software for a 32 K microcomputer of the times (Tall, 1986a, 1986b). The approach was totally graphic; no symbolic manipulation was involved. The software covered differential and integral calculus and differential equations. The following reports on the differential calculus, in particular, the use of local straightness, referred to above.

The dominant concept behind the computer programs was using the software as a generic organizer. It was generic in that the software directed the learner's attention to general features of the derivative through examples and counter-examples; it was an organizer in a similar sense to that used by Ausubel *et al.* (1968) in speaking of an 'advance organizer' – an agent that directs the learner to salient features and away from misleading features.

The differential calculus software focused on local and global aspects of functions and their derivatives. Locally it magnified graphs so that differentiable functions appeared locally straight at a point, and functions that were not differentiable at a point, e.g. $f(x) = \mathrm{abs}(x)$ at $x = 0$, were seen never to be locally straight. Globally the program drew the gradient at a number of points for different functions and displayed the overall gradient function – the derivative.

Classroom research used three experimental classes of 16-year-olds in England (using the software) and five control classes (using a traditional approach). Each class had between nine and eighteen pupils. The groups were roughly comparable in terms of ability and the same material was covered by all classes.

In one post-test (scored out of 20 marks), students were asked to draw the derivatives of four graphs. The results are shown in the table.

Group	No. of students	Mean	Standard deviation
Experimental	43	16.9	4.2
Control	69	7.5	5.8

In another post-test item, students were shown the graph in Figure 12.2 and told that it is the derivative of one of the graphs in Figure 12.3. Of the 43 experimental students, 32 chose correctly, as opposed to 20 of the 69 control students.

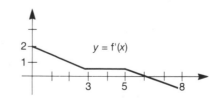

Figure 12.2

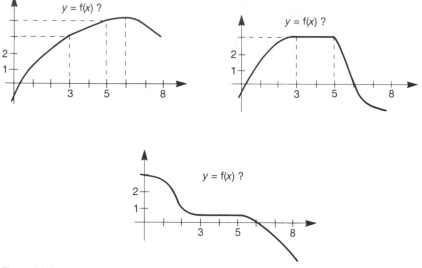

Figure 12.3

In a further post-test task, students were asked to give an example of a function defined at $x = 1$ but not differentiable there. Of the 43 experimental students, 15 answered correctly, as opposed to 1 of the 69 control students.

Tall concluded that this method of learning can enhance the geometric understanding of students of all abilities without adversely affecting their skills in formal manipulation.

It is interesting to note that Tall's software is different from either supercalculators or CASs in that it is designed as an educational tool, whereas supercalculators and CASs are mathematical assistants – they perform mathematical tasks; they are not designed with learning aims in mind. In this sense they are similar to basic calculators, which were designed for the business, not the educational, community.

Supercalculators

Ruthven (1990) reports on data collected at the end of the first year of a two-year project on using graphic calculators with 16–19-year-old students in England. The project culminated in the production of curriculum materials for teachers (National Council for Educational Technology, 1992). It enabled a number of teachers to use class sets of graphic calculators with at least one of their classes. The students owned the graphic calculators for this period. The teachers had no prior experience of using graphic calculators and were free to use them as they saw fit. Ruthven (1990) reports on the test results given to these students, the project group, at the end of their first year. A comparable set of students, the non-project group, who did not have access to graphic calculators also completed the test items, which covered material in two areas of the 16–19 mathematics curriculum where graphs are normally used. The project group were allowed to use their graphic calculators in the test.

Six questions, of which Figure 12.4 is an example, displayed graphs and requested an

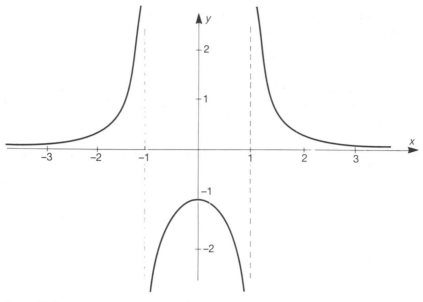

Figure 12.4

algebraic description. These are called the symbolization items. Other test items required students to interpret verbal questions in the context of distance–time and rate–time graphs. These were the interpretation items.

Ruthven's report concentrates on the symbolization items, because a number of distinctive approaches and important differences are suggested by the use of graphic calculators. Ruthven isolates three approaches to these items:

1. The *analytic-construction* approach. Here the students, in a similar way to the mature mathematician, exploit mathematical knowledge such as overall shape, points where f(x) = 0, asymptotes, etc. This was the most common method that students of both groups used, but the project group were able to check their answers by keying in the expected equation.

2. The *graphic-trial* approach. Here students use the graphic facilities in their calculators to modify symbolic expressions keyed in. Students may make a number of modifications before arriving at a suitable symbolic expression. Students are clearly translating from symbolic to graphic form, and vice versa, but this approach 'can be built on less extensive or confident knowledge' (Ruthven, 1990). About one quarter of the students in the project group used it. Although Ruthven's discovery is potentially very important, the phenomenon needs further research to clarify the nature of the process. Ruthven states that there are cases of a transition from the graphic-trial approach to the analytic-construction approach, and vice versa, but this does not answer doubts as to whether some students may get locked into the graphic-trial approach.

3. The *numeric-trial* approach. This has similarities to the graphic-trial approach, but here students try out numeric values and modify their original expression accordingly. This approach is generally not successful for functions other than linear

functions. About one third of the non-project group used this approach, but only one student in the project group did so.

These three approaches were not disjoint; there was overlap and, as mentioned, evidence of transition from one to another. The graphic-trial approach has major positive implications for widening access to higher mathematics to students in this age range. There is currently, in Britain, a shortage of students studying mathematics at higher levels. One reason for this is the difficulties students have in acquiring the skills and concepts that are tested at 16 and 18 years of age. It is possible that methods such as the graphic-trial approach could lead to more students being successful in mathematics.

The table displays the mean percentage score for the six symbolization items for both groups with respect to gender (group sizes in brackets). This mean score is adjusted for the grade that students obtained in the national mathematics examination at 16 years of age.

Group	Female	Male	All
Non-project	23 (14)	32 (19)	28 (33)
Project	63 (18)	52 (29)	57 (47)

The results are significant at two levels. The project group performed a great deal better than the non-project. Female students performed less well than males in the non-project group but this situation is reversed in the project group. Ruthven suggests that the higher performance by the project group is due to regular use of graphic calculators strengthening the relationship between graphic and symbolic forms and being a motivating factor for students and teachers. The gender effect is explained in terms of reducing anxiety and increasing confidence. Girls are generally more anxious and less certain of their abilities than boys. Personal technology allows them both to make their mistakes in private and to get immediate feedback.

Computer algebra systems

Heid (1988) reports on data collected from a 15-week introductory applied calculus course. The work was done at an American university. In America, unlike England and many other Western countries, calculus is an option in pre-university high-school courses, and introductory courses in calculus are similar to British A-level courses. Heid taught an experimental group of 39 students (in two classes of roughly equal numbers) using the CAS *MuMath*, a precursor of *Derive*. The control group consisted of 100 students on the same course who were taught together by traditional methods.

The control group were mainly taught the traditional calculus algorithms (the chain rule, product rule, etc., for differentiation, and various integration techniques). The experimental group used a CAS to perform most of these algorithms for the first 12 weeks of the course. The course concentrated on 'executive decision making in problem solving rather than on the actual execution of standard computational procedures'

(Heid, 1988). During the last 3 weeks of the course Heid taught the experimental groups the pencil-and-paper methods of performing the traditional algorithms.

Both the experimental and control group sat the same pencil-and-paper final examination, and a pencil-and-paper conceptual comparison test at the end of week 12, designed to examine non-algorithmic understanding of calculus concepts. Heid also used a range of other data-collection techniques – interviews and questionnaires throughout the course, quizzes, classroom tapes and photocopies of student assignments.

The students in the experimental class, according to Heid (1988), 'showed better understanding of course concepts'. Evidence for this claim came first from interviews, where students in the experimental class showed a greater ability to reason from basic principles and were generally able to reconstruct concepts for themselves. Students in the control group, however, rarely reasoned from basic principles and 'often alluded to having been taught the relevant material but being unable to recall what had been said in class' (Heid, 1988). Interview evidence for Heid's claim was further supported by the conceptual comparison test. The experimental group outscored the control group on all but two of the sixteen conceptual questions; students in the control group performed better on matching the graph of a function with a formula for its derivative. Students in the experimental group 'were better able to draw conclusions about slopes, identify portions of graphs reflecting given quantitative statements, [and] translate mathematical statements into conclusions about an applied situation' (Heid, 1988).

In the final examination, the experimental group performed almost as well as the control group (the differences in every section except 'optimize f(x, y)' were negligible). This result is very important, for it shows that even without a revised calculus curriculum, it is possible to resequence the skills element in a calculus course.

Subsequent to Heid's study, Palmiter (1991) conducted a study with a group of students very similar to Heid's sample. Seventy-eight students, evenly divided into control and experimental classes, took an identical conceptual and computational calculus test at the end of a period of instruction. The experimental group were able to use their CAS, which was MACSYMA, in the test but were penalized for careless errors. This group finished the course in half the time the control group took and were given one hour for the tests, whereas the control group were given two hours. The table shows the test results as means (standard deviation in brackets). The results clearly show a superior performance in the test scores for the CAS group. Although these results are significant at the 0.1 per cent level, using Hotelling's T^2 test, the difference in the examination conditions must be kept in mind, since the CAS group's course was clearly designed to enhance conceptual understanding and they had access to a CAS for the computational test.

	Classes	
Examination	CAS	Traditional
Conceptual	89.8 (15.9)	72.0 (21.4)
Computational	90.0 (13.3)	69.6 (24.2)

Course evaluation forms revealed that 85 per cent of the CAS group, as opposed to 68 per cent of the traditional group, felt confident about continuing the calculus course. Moreover, 73 per cent of the CAS group, as opposed to 43 per cent of the traditional group, felt that they had learnt more than in their previous mathematics courses.

Monaghan and Etchells (1993) reports on a number of small-scale classroom studies using CASs by Leeds University Master of Science students. In these, Etchells reports on 16–17-year-old students' work on integration. This study indicates that students using a CAS have a better understanding of integration as an infinite summation than do students following a traditional course, and that they use this understanding to reason correctly when integration is applied to new areas in mathematics. An analysis of videotapes of students working with CASs, moreover, revealed some interesting features. One student who was very able was not at all comfortable using a CAS. Also CASs can deceive the teacher in that students can appear to be producing intelligent mathematics when they are generating algebraic gibberish.

Hurd (in Monaghan and Etchells, 1993) examined 16+-year-old students studying the Newton–Raphson iterative method for finding the approximate root of an equation, i.e.

$$x_{n+1} = x_n - \mathrm{f}(x_n)/\mathrm{f}'(x_n).$$

Two small groups received four hours' tuition. One group used a spreadsheet, the other a CAS. The spreadsheet group performed better on the test items and appeared to understand the process more fully than the CAS students. Hurd suggests the reason for this is that in doing so much of the work for the students – even doing the differentiation internally – the CAS obscures the process. It is a black box. The spreadsheet work required a greater input from the students and displayed results that were closer to pencil-and-paper ones. If this is correct, then it is important that more work along these lines is carried out to determine what medium, including pencil and paper, best fits different topic areas.

CONCLUSION

The character of mathematics is changing, with developments in new technology progressing at a rapid pace. Mathematics teachers at all levels need to respond to the changes, and this is especially true for upper secondary teachers because the mathematics they teach is closer in content to the higher-level mathematics that is changing.

The upper secondary mathematics curriculum is a suitable vehicle to accommodate these changes, but the mathematics education community will have to address the place of old algorithms and new algorithmic procedures. If this is not done, then the problems of the slow progress of the electronic calculator will be revisited. The changes are certainly going to come. Will the mathematics education community respond proactively or reactively?

Research studies generally support the thesis that new technology can improve learning, but further research is needed. These studies need to examine how students learn, what resources are required and how these resources should be arranged.

REFERENCES

Atiyah, M. F. (1984) Mathematics and the computer revolution. *Nuovo Civilita Macchine (Bologna)* **2**(3), reprinted in A. G. Howson and J.-P. Kahane (eds) (1986), *The Influence of Computers and Informatics on Mathematics and its Teaching*. Cambridge: Cambridge University Press.

ATM (1982) *Function Graph Plotter* in Slimwam 1 (BBC compatible software). Derby: ATM.

Ausubel, D. P., Novak, J. D. and Hanesian, H. (1968) *Educational Psychology: A Cognitive View*. New York: Holt, Rinehart and Winston.

Bell, A. W., Costello, J. and Küchemann, D. E. (1983) *A Review of Research in Mathematical Education. Part A: Learning and Teaching*. Windsor: NFER.

Bibby, N. (1991) Wherefore 'plug and chug'? *Mathematical Gazette* **75**, 40–8.

Brayne, P. R. (1988) *Omnigraph*. Leamington Spa: Software Production Associates.

Cockcroft, W. H. (1982) *Mathematics Counts*. London: HMSO.

Corlett, P. (1985) *Introduction to Calculus*. Faringdon: Independent Software.

Cornu, B. (1988) The computer: some changes in mathematics teaching and learning. In J. de Lange and M. Doorman (eds), *Senior Secondary School Mathematics Education*. Utrecht: OW & OC.

Cornu, B. (1991) Limits. In D. O. Tall (ed.), *Advanced Mathematical Thinking*. Dordrecht: Kluwer.

Costello, J. (1992) A failed revolution. *Micromath* **8**(1), 21–3.

DES (1991) *Mathematics in the National Curriculum*. London: HMSO.

Dubinsky, E. and Tall, D. O. (1991) Advanced mathematical thinking and the computer. In D. O. Tall (ed.), *Advanced Mathematical Thinking*. Dordrecht: Kluwer.

Etchells, T. A. (1993a) Computer algebra systems and students' understanding of the Riemann integral. In J. Monaghan and T. A. Etchells (eds), *Computer Algebra Systems in the Classroom*. Leeds: University of Leeds Centre for Studies in Science and Mathematics Education.

Etchells, T. A. (1993b) The fall and rise of the chain rule. *Mathematical Gazette* **77** (in press).

Harper, D., Wooff, C., and Hodgkinson, D. (1991) *A Guide to Computer Algebra Systems*. Chichester: John Wiley.

Heid, M. K. (1988) Resequencing skills and concepts in applied calculus. *Journal for Research in Mathematics Education* **19**(1), 3–25.

Howson, A. G. and Kahane, J.-P. (eds) (1986) *The Influence of Computers and Informatics on Mathematics and its Teaching*. Cambridge: Cambridge University Press.

Hunter, M. and Monaghan, J. (1993) School algebra and computer algebra systems. *Micromath* **9**(3) (in press).

Lyons, J. L. (1988) Innovation in calculus textbooks. In L. A. Steen (ed.), *Calculus for a New Century*. MAA Notes No. 8. Mathematical Association of America.

MEI (1983) *Programs for Mathematical Computing – I*. Peterborough: MEI.

Monaghan, J. and Etchells, T. A. (eds) (1993) *Computer Algebra Systems in the Classroom*. Leeds: University of Leeds Centre for Studies in Science and Mathematics Education.

National Council for Educational Technology (1992) *Personal Technology in the Classroom*. Coventry: NCET.

Palmiter, J. R. (1991) Effects of computer algebra systems on concept and skill acquisition in calculus. *Journal for Research in Mathematics Education* **22**(2), 151–6.

Reid, B. W. (1990) *Mathematics and the Graphic Calculator*. Oxford: Oxfordshire Mathematics Centre.

Ruthven, K. (1990) The influence of graphic calculator use on translation from graphic to symbolic forms. *Educational Studies in Mathematics* **21**(5), 431–50.

Schools Council (1972) *From Counting to Calculating: A Study of Arithmetic for Secondary Pupils*. London: Chatto and Windus.

Shuard, H., Walsh, A., Goodwin, J. and Worcester, V. (1991) *Calculators, Children and Mathematics*. Hemel Hempstead: Simon & Schuster.

SCUE and CNAA (1978) *A Minimal Core Syllabus for A-level Mathematics*. Watford: Edson.

Shumway, R. (1988) Graphic calculators: skill versus concepts. In J. de Lange and M. Doorman (eds), *Senior Secondary School Mathematics Education*. Utrecht: OW & OC.

Skemp, R. R. (1977) Relational and instrumental understanding. *Mathematics Teaching* 77, 20–26.

SMP 16–19 (1991) *Introductory Calculus*. Cambridge: Cambridge University Press.

Tall, D. O. (1985) *Supergraph* (BBC compatible software). London: Glentop Press.

Tall, D. O. (1986a) *Graphic Calculus I, II, III* (BBC compatible software). London: Glentop Press.

Tall, D. O. (1986b) Building and testing a cognitive approach to calculus using interactive computer graphics. Unpublished Ph.D. thesis, University of Warwick.

Tall, D. O. (1990) The transition to advanced mathematical thinking; functions, limits, infinity and proof. In *The National Council of Teachers of Mathematics Handbook on Research in Mathematics Education*. Reston, VA: NCTM.

Tall, D. O. (ed.) (1991) *Advanced Mathematical Thinking*. Dordrecht: Kluwer.

Tall, D. O. and West, B. (1986) Graphic insight into calculus and differential equations. In A. G. Howson and J.-P. Kahane (eds), *The Influence of Computers and Informatics on Mathematics and its Teaching*. Cambridge: Cambridge University Press.

Chapter 13

Postscript: The Future of Mathematical Education

Tom Roper

Previous chapters of this book have each dealt with an issue of current interest within mathematical education. The purpose of this chapter is to look at how some of these issues might interact with each other in the near future. Such an undertaking is a risky one, since the future very quickly becomes the present and then the past. Nevertheless, unless we have some conception of the future pressures upon mathematical education and where they may push us, then we are more than likely to drift into situations over which we will have a decreasing amount of control.

Perhaps the most obvious pressures upon mathematical education both now, and for the foreseeable future, are the developments within information technology relating to both hardware and software (see Chapter 12). Whatever can be calculated can be computed, and since the majority of mathematics lessons currently taught in most classrooms throughout the world consist of how to do certain calculations, or algorithms as they are now termed, the impact of this technology could be devastating in its effect, or liberating beyond all measure. It could be devastating in that, if there are machines capable of doing virtually all the mathematics that the average citizen might need, is there then any need for universal education in mathematics other than basic arithmetic? It could be liberating in that any individual will have access to a whole armoury of mathematical techniques, which, set in the context of a mathematical education that acknowledges this, confers great power and independence upon the individual.

The key issues that this technology forces us to face are why we want our children to learn mathematics and what we want them to learn from it. The answers to these questions clearly must be the basis of our aims for mathematical education (see Chapter 1), and cannot be avoided. Handheld supercalculators, or palmtop computers, will soon be available at a price comparable to that of a graphic calculator at present. The economics of the market place will ensure that this will happen. Therefore a 13- or 14-year-old pupil of any ability will have access to one, through either a set provided by the school or personal ownership, the latter being the more likely.

The possession of personal technology by pupils will force us to face this challenge of defining our aims. The pupils themselves will raise the issue by demanding to know why they have to do things in certain ways when it is obvious that they have in their possession a machine, or some equivalent piece of equipment, which will do it much more quickly, with greater accuracy and more directly. I witnessed at first hand several years ago the strength of feeling which such an issue can generate. A trainee teacher on teaching practice had been given the task of teaching logarithms to a middle-ability class of 13- and 14-year-olds. This was perhaps the last year that children of this age were taught logarithms before revision of syllabuses made them redundant. The pupils were trying to come to grips with multiplication and division involving numbers less than unity. One boy had a calculator and asked repeatedly if he could use it. The teacher said he could not; logarithms were a part of the syllabus and he had to learn how to use them. Variants upon this theme mingled with patently false claims concerning the practicality of logarithms as a means of doing calculations finally prompted the pupil, after yet one more justification from the teacher, to cry out at the top of his voice, 'Yes, but why?' The frustration was obvious. The trainee teacher stood aghast, with no available reply, and the pupil proceeded to do the questions with his calculator. We must be able to answer the question of why, to the satisfaction of our pupils. If we cannot, then they will reject the mathematics which we offer them just as surely as the pupil whom I observed did.

Intertwined with this is the issue of how children learn and what it is that they need, and are able, to learn. My 7-year-old son wanted to know how many hours there are in a year. We soon established that there were 24 hours in the day and 365 days in the year; he was prepared to ignore the complication of leap years. He then asked for my calculator and used it to perform the necessary computation without any further assistance. If he knows when to multiply and knows his basic multiplication tables, does he really need to know how to do long multiplication? Similarly, if it is known that the equation $a.b = 0$ implies that either $a = 0$ or $b = 0$ or both, is it actually necessary to be able to factorize polynomials when a symbol manipulator can do it?

Both of these are examples of algorithms currently taught at various stages in the process of mathematical education. If we do not teach them or give them sufficient emphasis, then we have to consider whether there might not be some form of penalty to be paid at a later stage. Thus Gardiner (1987) argues for the teaching of the long division algorithm because it prepares the way for the division of polynomials and the remainder theorem in the algebra required of A-level students. This is but one example among several which Gardiner gives. Doubtless there are still others. Yet the pace of technological development may force us into hasty decisions which could result in the severing of some of these important connections. Bibby (1991) says that 'We can see microtechnology driving a wedge between mathematical content and the associated traditional mathematical skills.' However, we cannot see precisely what this wedge is cutting through, the 'mathematical cement' which holds content and skills together. Do we wish to preserve this 'mathematical cement', and if so how do we go about it? There is a clear need for research into the use of symbol manipulators and other forms of information technology in the classroom to establish modes of use, what pupils learn from using them and how they learn with them. Without answers to these questions, we are in danger of making decisions based upon wishful thinking which we may have cause to regret in the future.

Of course, much of this debate focuses upon our relationship, as trained mathematicians, with mathematics, and upon how our experiences have formed that relationship. It would be wrong to expect that our pupils will form the same kinds of relationships with the subject in the same way, since their experiences are, and will be, very different from ours. Indeed, the constructivist philosophy (see Chapter 3) would say that all our experiences are open to individual interpretation and that none of us acquires the same understandings from the same learning situations as anyone else. Therefore there are potential dangers in projecting our relationship with mathematics and what we see in the technology onto our pupils. The gravest of these may well be the heightened expectations we might have that our pupils will be able to construct, from the outputs of a machine, the same understanding that we possess but have attained via an entirely different route. We must expect something different, and not just in terms of maturity.

The argument propounded in favour of the calculator was that by removing the drudgery of calculation, pupils would be able to concentrate on solving problems from the world outside the classroom. Similar arguments are put forward for the syllabuses of the future based upon the symbol manipulator. The *DERIVE User Manual* (Rich *et al.*, 1992) comments, 'This [elimination of the drudgery of performing long tedious mathematical calculations] gives you the freedom to explore different approaches to problems – approaches that you probably would not consider if you had to do the calculations by hand.' This explicitly promises new approaches, and therefore potentially different understandings, within an exploratory framework. Implicitly this commits us to a modelling approach similar to that described in Chapter 11, wherein each pupil can formulate his or her own model of the problem and its relationships, trying them out against the reality of the problem.

The teaching of mathematics through the solving of problems is the paradigm which is often seen as best fitting with the constructivist philosophy of how mathematics is learned. However, current obsessions with comparisons of mathematical performance at international, national, local area, school and individual level mean that teachers will concentrate upon what they see as the efficient attainment, by their pupils, of the behavioural objectives of assessment in their teaching. Not for the first time, there is a tension between the assessment of mathematics and the theories of how mathematics is learned. The hope that technology brings with it is that in removing the drudgery of calculation – that is, the majority of the behavioural objectives – assessment may be able to offer more constructivist-oriented models.

An example of this is provided by the graphic calculator. If these machines are permitted within the examination room, then it is pointless to ask the candidate to sketch the curve of a given equation. However, candidates can be asked to identify the family of a particular curve from a sketch, or, given sufficient information, suggest possible forms for its equation (see Chapter 12). Thus, by constructing the answer, candidates are reconstructing their own knowledge, building a model and testing it against the available information. The symbol manipulator will offer many similar opportunities in the future over a much wider field. However, until such assessments become standard, the assessment of the processes of mathematics, problem solving, modelling and the like will be tied to, and therefore subordinate to, the assessment of mathematical content as represented by the performance of standard algorithms and solving standard problems.

Technology as seen in the microcomputer is expensive. It can be used effectively by no more than two to three people working in consort. On the other hand, one microcomputer per classroom for class teaching and individual or group-based work seems to be the recommended minimum. By comparison, the personal computer embodied in the palmtop is cheap and, while unsuitable for class teaching, may be used effectively by two people. These may seem relatively unimportant facts. However, in terms of the budget of developing countries, they are quite vital. By and large, the mathematics syllabuses of developing countries are derived from those of 'Western' countries. Because curriculum development has been moderately uniform in global terms, there are many similarities between countries in the mathematics that is taught. A technologically driven revolution in the mathematics curricula of the future could leave those countries that cannot afford the necessary quantities of hardware for the classroom with either outdated syllabuses or the prospect of inflicting unnecessary burdens of calculation upon their students in an effort to teach the new mathematics, whatever it may be. The miniaturization of technology therefore provides an opportunity for the developing world to purchase the necessary hardware, perhaps not in sufficient quantities to permit ideal classroom provision, but hopefully sufficient to enable it to stay in touch with the developed world.

So far a great deal of hoped-for change has been invested in the development of microchip technology. This is because it is the one independent element in all that has been considered. Developments within mathematics itself have fuelled changes within the school curriculum in the past. Such developments were one of the driving forces behind the so-called modern mathematics revolution. However, the latest developments within mathematics are well and truly founded upon the computer. Without the aid of technology, such topics as iteration, chaos and fractals are impossible to contemplate (see Chapter 6). On the other hand, some topics in discrete mathematics might appear, at first glance, to be not as dependent upon technology (see Chapter 5). However, a closer look at those syllabuses that have been developed so far shows that the mathematics within them is applied to problems which either seem to be obvious and so not to merit the use of the mathematics at all, or involve considerable tedious calculation. If problems which truly justify the mathematics being taught are to be introduced, then the use of appropriate technology is essential.

Syllabuses are now almost universally controlled by the state and not by the teaching profession. They change slowly, under internal pressure, with difficulty. There can therefore be no independent development of syllabus unless there is sufficient pressure caused by an independent and external source. That source, as argued above, can only be technology.

However, the teaching profession, to a large extent, still has control over how it teaches – the methods which it employs. It is perhaps in this area that teachers may look for some degree of autonomy, in trying to develop different teaching styles to accommodate different learning styles, and different contexts in which to place the content that it is incumbent upon them to teach (see Chapters 9 and 10). Certainly, the United Kingdom, via the National Curriculum of England and Wales (NC), seems to be the only country which has made the process aspects of mathematics compulsory. There is, though, considerable international interest in problem solving as a field of research and as a teaching paradigm. This interest may mark the spread of more diverse modes of operation in the mathematics classroom.

There is also growing international interest in basing the content in the context of the pupils' culture. This poses considerable difficulty, not just in terms of 'unfreezing' the mathematics frozen within the culture (see Chapter 2) but in having that mathematics accepted as being legitimate. This is especially true where education is seen as advancement away from, or out of, one's own culture. In such circumstances, to use the culture as context could be interpreted as attempting to lock pupils into their particular culture, leading to a rejection of mathematics.

The mother tongue of a pupil is an important factor in the pupil's learning mathematics. Many developing countries have the early years of compulsory education in the mother tongue and then switch to the language of their former colonizers (see Chapter 7). However, the problem of language is not confined to the developing world; migration is bringing this problem into the classrooms of the developed world too. In several of the United Kingdom's inner city schools, pupils of considerable mathematical ability are trapped in the lower sets, suffering a course which does not meet their intellectual needs, until either their language skills improve sufficiently or they are provided with access to the curriculum via language support. Recognition that this is an international problem may help to increase the amount of research done in this field and also generate guidelines for the production of classroom materials.

Whatever the future holds for mathematics education, there is one certainty: that the classroom teacher will bear the brunt of any changes and will ultimately be responsible for seeing that they are carried out effectively. This places the education of teachers as the major priority for the future, in terms both of initial training and support, and of development throughout their careers. Training in post can only be effective if teachers are receptive to change, and have some feel for it from the day they enter the profession. This implies that the initial training of teachers must be for the future, not for the immediate present. Attempts to concentrate the training of teachers within the classroom when many classroom teachers look outside for help in meeting change seems to be contradictory. If the general quality of mathematics education is to be improved, it will be so most directly and immediately from a well-qualified and well-trained teaching force which is informed about the issues and prepared for change, not from a microchip or curriculum reform.

REFERENCES

Bibby, N. (1991) Wherefore 'plug-and-chug'? *Mathematical Gazette* **75**, 40–8.
Cockcroft, W. H. (1982) *Mathematics Counts*. London: HMSO.
Gardiner, A. (1987) Fishy chips. *Mathematics in School* **16**(5), 16–17.
Rich, A., Rich, J. and Stoutemyer, D. (1992) *DERIVE User Manual. A Mathematical Assistant for Your Personal Computer*. Honolulu: Soft Warehouse, Inc.

Name Index

Subject Index

Several of the entries in the index are given as abbreviations. A list of the abbreviations used in the book can be found on page xii.